Sustainable Aviation

The Sustainable Aviation book series focuses on sustainability in aviation, considering all aspects of the field. The books are developed in partnership with the International Sustainable Aviation Research Society (SARES). They include contributed volumes comprising select contributions to international symposiums and conferences, monographs, and professional books focused on all aspects of sustainable aviation. The series aims at publishing state-of-the-art research and development in areas including, but not limited to:

- Green and renewable energy resources and aviation technologies
- Aircraft engine, control systems, production, storage, efficiency, and planning
- Exploring the potential of integrating renewables within airports
- Sustainable infrastructure development under a changing climate
- Training and awareness facilities with aviation sector and social levels
- Teaching and professional development in renewable energy technologies and sustainability

Selim Gürgen
Editor

Failure in Aircraft Materials

Springer

SARES
INTERNATIONAL SUSTAINABLE AVIATION
AND ENERGY RESEARCH SOCIETY

Editor
Selim Gürgen
Department of Aeronautical Engineering
Eskişehir Osmangazi University
Eskişehir, Türkiye

Aerospace Technologies Research Center
Eskişehir Osmangazi University
Eskişehir, Türkiye

ISSN 2730-7778 ISSN 2730-7786 (electronic)
Sustainable Aviation
ISBN 978-3-031-65849-5 ISBN 978-3-031-65850-1 (eBook)
https://doi.org/10.1007/978-3-031-65850-1

This Springer imprint is published by the registered company Springer Nature Switzerland AG
The registered company address is: Gewerbestrasse 11, 6330 Cham, Switzerland

Preface

Material reliability is one of key factors in flight safety. For this reason, engineers and material scientists make a lot of effort to enhance the strength of materials to cope with harsh conditions during flights. Despite this hard work, aircraft sometimes confront with material failures. The book *Failure in Aircraft Materials* discusses the fundamental and the most frequent failure modes in aircraft alloys. Root cause of failures is discussed as well as presenting case studies from the sector. Theoretical knowledge is brought together with applications, thereby providing a fruitful source for scientists, researchers, and engineers in aerospace industry.

Eskişehir, Türkiye | Selim Gürgen

Contents

About the Editor

Selim Gürgen is an Associate Professor with the Department of Aeronautical Engineering at Eskişehir Osmangazi University, and the Director of Eskişehir Vocational School. He is an Executive Board Member of the ESOGU Aerospace Technologies Research Center. He was a visiting researcher at the University of Wollongong for a project on smart materials. He was at the University of Aveiro on natural composites, and worked as a researcher with the School of Transportation at Anadolu University for four years. Before his academic life, he worked as a manufacturing engineer in a joint venture of KMWE-Dutch Aero. Dr. Gürgen has contributed to several industrial and research projects in collaboration with major organizations such as Fiat Automobiles, Turkish Aerospace Industries (TAI), and the Turkish Air Force. His projects were funded by the Scientific and Technological Research Council of Turkey (TÜBİTAK) and the European Union (EU). He has many publications in refereed international journals and conferences. He serves as an editorial member for several international journals.

Chapter 1
Solid Lubricants Against Wear Formation in Aerospace Components

Gökçe Mehmet Ay

1.1 Introduction

Objects that are in motion and in contact with each other cause friction and wear. These effects have been known since ancient Egyptian times. Lubrication is one of the important mechanisms designers and engineers use to cope with the effects of wear. It was estimated that energy loss via friction and wear is nearly 20% of global energy consumption every year [1]. When you add the demand for high performance with the technological development, the requirements for lubrication increase. An important area for lubrication is the aerospace industry, with its extreme conditions.

Aerospace components have high reliability requirements. Those components work under high or low temperatures, at high speeds and pressures, or in the case of space vehicles vacuum and high radiation. These components should operate under such conditions with no failure. However, moving parts are prone to failure because of wear.

Some of the moving components in aerospace applications are:

- Rocket engines.
- Couplings.
- Bearings.
- Gimbals, gyroscopes, and other aerospace instrumentation.
- Gyroscopes.
- Bearings for thrust and vector control.
- Seals for valves and pumps.
- Space suit articulation points and joints [2].

G. M. Ay (✉)
Department of Mechanical Engineering, Eskişehir Osmangazi University, Eskişehir, Turkey
e-mail: gma@ogu.edu.tr

S. Gürgen (ed.), *Failure in Aircraft Materials*, Sustainable Aviation,
https://doi.org/10.1007/978-3-031-65850-1_1

Materials that fight friction, or "tribological materials," require more than just wear resistance. They also must be strong, stiff, and able to handle heat and stress. Making all these things work together in one material is a tough design challenge.

Friction and wear are often reduced by inserting a shear-accommodating layer (such as a thin film of liquid) between contacting surfaces. When working circumstances exceed the limits of liquids (for example, high temperatures or vacuums that are present in aerospace applications), or when liquids cannot be introduced, attention shifts to solid lubricant coatings.

The solid lubricating action of naturally occurring minerals like graphite and molybdenite is well understood. However, the advancement of current coating processes, such as physical vapor, chemical vapor, and plasma-enhanced chemical vapor depositions, has significantly increased the use of solid lubricating materials [3].

Types of solid lubricants used in aerospace applications are:

- Nanotubes, nano-onions, and other nanoparticles (C, BN, MoS_2, and WS_2).
- Nanocomposite coatings (WC/C, MoS_2/C, WS_2/C, TiC/C, and nanodiamond).
- Diamond and diamond-like carbon (DLC) coatings (diamond, hydrogenated carbon (a-C:H), amorphous carbon (a-C), carbon nitride (C_3N_4), and boron nitride (BN) films).
- Superhard or hard coatings (VC, $B4_C$, Al_2O_3, SiC, Si_3O_4, TiC, TiN, TiCN, AlN, and BN).
- Lamellar film (MoS_2 and graphite).
- Nonmetallic film (titanium dioxide, calcium fluoride, glasses, lead oxide, zinc oxide, and tin oxide).
- Soft metallic film (lead, gold, silver, indium, copper, and zinc).
- Self-lubricating composites (nanotubes, polymer, metal-lamellar solid, carbon, graphite, ceramic, and cermets).
- Lamellar carbon compound film (fluorinated graphite and graphite fluoride) Carbon Polymers (PTFE, nylon, and polyethylene).
- Fats, soap, wax (stearic acid), ceramics, and cermets [4].

The most important benefit of solid lubrication is the ability to maintain lubrication in inaccessible or harder-to-reach areas of an aircraft or spacecraft. This property of solid lubricants plays a crucial role in enhancing the reliability of components. Solid lubricants are used in aircraft engines, turbine blades and landing gear in order to prevent fretting wear and corrosion [5]. Another benefit of solid lubricants over liquid lubrication is its ability to be stored without losing lubrication properties. For example, aircraft equipment can be stored on a shelf for a long time, or a spacecraft component such as an antenna can remain stationary for a long period. Solid lubricants are very effective in environments with high dust content. The dust in the environment generates a paste with liquid lubricants and thus a grinding effect is observed. Solid lubricants, such as diamond-like coatings, protect the equipment from abrasive wear caused by dust particles.

In this chapter, the nanoparticles, graphite, molybdenum disulfide (MoS_2), tungsten disulfide (WS_2), and hexagonal boron nitride (hBN), are discussed. The lubrication mechanisms of these solid lubricants and best practices for wear control are also discussed.

1.2 Solid Lubricant Properties and Lubrication Mechanisms

Solid lubricants can be used on tribological surfaces through several methods. The traditional approach involves dispersing or rubbing small granules of solid lubricant onto the surfaces. Another effective method involves using fine powders of various solid lubricants to lubricate sliding bearing surfaces. Some solid lubricants are combined with an aerosol carrier and sprayed directly onto the surfaces needing lubrication. Alternatively, solid lubricant powders can be firmly adhered to a surface using appropriate adhesives or epoxy resins, enhancing wear resistance. They can also be dispersed or injected into composite materials. However, in many current applications, thin layers of solid lubricants are preferred over powdered or bonded formulations [6].

Layered materials represent a significant category of solid lubricants, encompassing substances like graphite, MoS_2, and hBN, all renowned for their lubricating properties. The structure of these materials is illustrated in Fig. 1.1. These

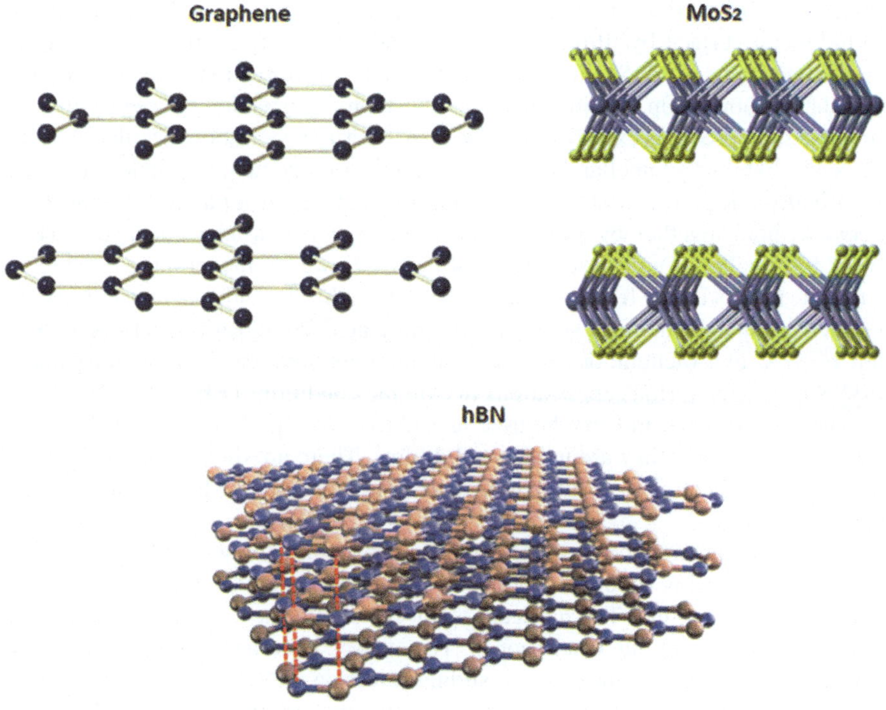

Fig. 1.1 Lamellar structures in solid lubricants [9, 10]

lubricating films offer the advantage of being solids, suitable for use in open systems and scenarios where liquid lubrication is unsuitable. Regardless of the application, they facilitate interlayer sliding interfaces with remarkably low interfacial shear strengths, owing to weak interlayer interactions within the materials [7]. Additionally, referred to as two-dimensional materials, these substances gained attention with the emergence of graphene as a solid lubricant. Graphene comprises a single layer of carbon atoms arranged in a two-dimensional honeycomb lattice. It serves as a fundamental component for various well-known carbon compounds, including three-dimensional graphite, one-dimensional carbon nanotubes, and zero-dimensional fullerene [8].

The lubricating properties of two-dimensional materials stem from their lamellar crystalline structure. For instance, graphite exhibits a hexagonal shape composed of planes of polycyclic carbon atoms. The bonding between these basal planes is relatively weaker due to the larger distance between carbon atoms along the c-axis, with parallel layers of hexagons spaced approximately 0.34 nm apart. Upon closely examining a metallic mating surface moving across graphite, coefficients of friction ranging from 0.05 to 0.15 in ambient conditions were observed. However, the coefficient of friction increases significantly, sometimes threefold or more, when sliding occurs perpendicular to the basal planes compared to sliding parallel to them [11].

Another layered solid lubricant, MoS_2, possesses a lamellar structure where the sulfur lamellae are connected only by weak interactions, typically van der Waals forces, which effectively reduce shear [12]. Upon application, MoS_2 crystals adhere to the metal surface by filling gaps or embedding into it. During movement, the crystalline layers of MoS_2 slide and align parallel to the direction of relative motion, facilitating lubrication. Additionally, the robust ionic connection between sulfur and molybdenum provides excellent resistance to asperity penetration. Similar to MoS_2, hBN also exhibits a lamellar structure. Its atomic arrangement resembles graphene, with boron and nitrogen atoms occupying the vertices of a planar hexagonal sp2 network, held together by van der Waals forces between adjacent layers. These weak forces allow for easy separation of layers. However, the stronger bond to the substrate results in low friction characteristics [13]. hBN stands out due to its exceptional thermal and chemical stability, exhibiting no decomposition up to 1000 °C in air, 1400 °C in a vacuum, or 2800 °C in an inert environment. This stability makes hBN suitable for various applications in extreme conditions [14].

These solid lubricants can be used as additives to liquid lubricants [15], but in extreme conditions, they are used as solid films. There are three main methods for aerospace applications of solid lubricants. These are burnishing, bonding, and sputtering.

The burnishing method is performed by either rubbing the solid lubricant into the desired surface using a cloth or a brush or with a volatile solvent that evaporates and leaves the solid behind. The lubricant is not bonded to the substrate surface but is held in place by van der Waals bonds. The surface coverage is greater, and solid lubricant crystals are aligned to the sliding direction. However, the weak van der Waals forces might cause poor adhesion, limiting the use of burnishing [16]. A proposed solution to this problem is pretreatment of surfaces with pressure pads [17].

Another solution is to use highly pressurized solid lubricant particles that are sprayed on substrate surfaces. This method is cost-effective, and it is possible to obtain thin coatings with less than 1 mm thickness [18]. The substrate should be adequately prepared and cleaned before application [19]. Although it is very important to clean the substrate according to the standards [20], surface roughness is vital as well. Burnished solid lubricant coatings perform better with some surface roughness [21].

Bonding solid lubricants generates surfaces with better characteristics. These characteristics are summarized in a study by Chen et al. [22].

- It is possible to apply bonded solid lubricating coatings, which are thin (several microns), to different tribological parts and components, including engine components, landing gears, and airframes, without altering their size or shape.
- Bonded solid lubrication coatings demonstrate excellent performance across a range of challenging environments, including high temperatures, ultra-low temperatures, ultra-high vacuum, strong radiation, strong redox conditions, and other extreme scenarios where traditional fluid lubricants are inadequate or environmentally harmful. For instance, these coatings are well-suited for application on high-temperature and high-pressure turbine blades, combustion chambers, or rocket nozzles, providing effective lubrication in demanding conditions.
- Bonded solid lubrication coatings have high load-bearing capacity and pressure resistance, which means they can withstand high forces and stresses without breaking or wearing out. A bonded solid lubrication coating with robust pressure resistance exhibits a load-bearing capacity that surpasses oils and greases by several times or even up to a hundredfold.
- Bonded solid lubrication coatings are ideal for lubricating mechanical components that experience frequent starts and intermittent operation, yet remain reliable during extended periods of inactivity, such as valves, switches, or sensors. Unlike fluid lubricants, these coatings do not deteriorate over time or dissipate. Moreover, they can be employed alongside oils and greases to enhance safety through emergency lubrication and optimize the running-in performance of heavily stressed components like gears, bearings, or pistons.
- Bonded solid lubrication coatings provide corrosion resistance, dynamic sealing, vibration reduction, noise reduction, and "stick-slip" prevention to the coated parts and components. Bonded solid lubrication coatings form a protective film on the substrate that prevents corrosion and leakage, reduces mechanical vibration and noise, and prevents the occurrence of "stick-slip," which is a phenomenon where the sliding surfaces alternately stick and slip, causing jerky motion and wear.
- Bonded solid lubrication coatings offer the capability to lubricate a wide array of materials, including metals, plastics, elastomers, ceramics, and glass. This is particularly advantageous as oils and greases often struggle to function effectively on such surfaces and may even cause damage.

The pretreatment is important for bonded coatings. Degreasing for nonmetallic components, sometimes followed by blasting, is recommended. Rust and scale on metals must be chemically removed using acid or lye pickling solutions. Additional treatments such as anodizing, plating, nitriding, chemical oxidation, vulcanizing, and phosphating may be required to increase the solid film lubricant's adherence to the metal surface. Curing and finishing are part of the posttreatment procedure for bonded solid lubricating coatings. Depending on the components of the bonded system, various curing procedures can be used, including ambient temperature curing, heating curing, light curing, electron beam curing, and water curing. Overall, heat-cured bonded solid lubricating coatings outperform room-temperature-cured coatings [23]. The bonding materials are varied; polymers such as PTFE [24] or polyimide [25, 26] can be used.

Sputter-deposited films of solid lubricants such as MoS_2 have been used on a variety of spacecraft mechanisms. The first coatings were conducted using an RF technique in 1971 [27]. Later, different techniques and materials were applied. The most used solid lubricant is MoS_2; however, the effects of additional materials such as metals [28] or oxide dopants [29] have been examined.

1.3 Solid Lubricant Applications in Aerospace Components

Microstructurally designed solid lubricants are critical for improving lubrication systems at low and high temperatures, as well as high vacuum, pressure, and chemical reactivity [30]. Ballistic missiles, hypersonic transportation, and other motion systems work in environments with extremely high working temperatures and require solid lubricants [31].

1.3.1 Aircraft

The aircraft landing gear plays a crucial role as it bears the entire weight of the aircraft during landing and ground operations. Attached to the aircraft's principal structural elements, it is susceptible to wear and corrosion during operation, which can lead to catastrophic consequences if not addressed [32]. Landing gear failure poses significant risks due to factors such as excessive wear, fracture, mechanical failure, and erosion. Wear on critical landing gear components can cause deformation and material removal, ultimately leading to diminished efficiency. To mitigate these risks, coatings can be applied to protect stressed structural parts and decelerate the corrosion process of metallic components [33]. In order to limit the effect of wear due to component contact, solid lubricants are used. The dust particles, high loads, and thermal fluctuations prevent the usage of liquid lubricants in landing gears. The solid lubricants most commonly used are MoS_2, WS_2, and diamond-like carbon (DLC). These solid lubricants are applied via coatings, specifically in the case of DLC, and bonded films. Solid lubricants are also used as additives to landing gear component greases.

The usage of solid lubricants reduces the wear on bearings, thus extending their lifespan and reducing maintenance costs. The ability of solid lubricants to withstand wider temperatures and harsh conditions enables a wider range of operating conditions. As always, lower friction causes less heat generation and decreases fuel costs.

Another important component that needs lubrication in an aircraft is the aircraft engine. For a vehicle traveling on land or sea, it is possible to repair it in case of a lubrication failure. However, for aircraft, there are few options in the case of lubrication failure. The aircraft should be able to land safely after the failure within 30 to 15 min [34]. In order to prevent failure, designers might choose membrane lubrication techniques or solid lubricant coatings. There are studies using MoS_2 and graphene for this failure scenario [35].

1.3.2 Spacecraft

In broad terms, spacecraft can be categorized into three groups: remote sensing (e.g., meteorological), communication, and navigation satellites. Additionally, the tribology of launch vehicles plays a crucial role in successfully deploying spacecraft into their designated orbits. Launch vehicles and spacecraft feature distinct release mechanisms that facilitate the separation of the spacecraft from the launch vehicle. Moreover, spacecraft are equipped with deployment mechanisms that enable subsystems, such as antenna dishes and solar panels, which are typically folded during launch to conserve space, to be opened once in orbit. These release or deployment mechanisms require the use of lubricants to maintain low friction (torque) for a limited number of cycles [36].

The space environment is marked by severe temperatures, vacuum, micrometeoroids, space debris, and huge changes caused by sunspot activity. These factors play a significant role in the design and development of spacecraft and aerospace systems. Surfaces exposed to these systems deteriorate due to the presence of atomic oxygen (AO). Aerospace structures require materials with high strength and stiffness to maintain mechanical qualities due to high phase temperatures. Initially, this was accomplished using composites containing a graphene mixture and epoxy resin as a curing agent, which include a variety of materials with respect to expected outcomes in application areas such as increased base material strength, increased temperature resistance, and improved material tribological behavior [37].

1.4 Future Trends and Challenges on Solid Lubricants

Although current solid lubricants for wear prevention work adequately, there are still a few areas where research is focused on improving. The most interesting development is on new materials, such as MXenes and transition metal dichalcogenides (TMDs) with customized structures. MXenes represent a prominent group of two-dimensional transition metal-based materials derived from carbides and nitrides. These materials exhibit a characteristic layered structure, with individual layers composed of transition metal atoms (M) covalently bonded to carbon and/or nitrogen (X). Each layer further features an uneven distribution of functional groups (T), such as hydroxyl (-OH), oxygen (-O), or fluorine (-F), chemisorbed onto the metallic surface. The general formula for MXenes can be expressed as Mn + 1XnTx, where n is an integer value representing the stoichiometry of the functional groups. As a result of this unique structure and diverse surface chemistry, MXenes exhibit an array of intriguing properties with immense potential in various technological applications [38]. According to recent studies, MXenes demonstrate a distinctive capacity to deliver sustainable lubricity regardless of thickness, thus distinguishing them from other lubricants based on two-dimensional materials [39]. Tribology research has recently made strides toward tailoring two-dimensional TMDs to improve their tribological characteristics. The idea of TMD metamaterials was presented in a recent study. These materials can be engineered to have particular edge-plane-bulk characteristics in order to obtain the desired tribological qualities [40]. The structure modulation of TMDs, including point, linear, and out-of-plane structures, which can be tailored, enables improvements to their tribological performance [41].

Another area of study on using solid lubricants in aircraft and aerospace vehicles is application techniques. Laser surface texturing (LST) is a promising technique for enhancing the properties of aerospace alloys, as it can improve wear and corrosion resistance, wetting behavior, and mechanical characteristics [42]. The introduction of the laser implantation technique has further advanced this field, making it possible to create reliable textured surfaces with micro-elevated features [43].

Solid lubricants are more and more combined with other technologies for better performance, such as self-healing surfaces. These are surfaces where microcapsules within lubricants enable them to fix minor damage and maintain functionality. Although still an emerging area, self-healing surfaces with solid lubricants, using self-repairing lubricant-grafted surfaces, is an interesting area of research for aerospace applications [44, 45].

1.5 A Case Study: The Role of a Solid Lubricant in Frictional Interactions

To investigate the effect of hBN solid lubrication, wear tests with the ball-on-disc configuration were conducted. Specimens made of AISI 4140 steel with a diameter of 30 mm were prepared metallographically. The wear tests were carried out using the CSM tribometer under a load of 2 N and a speed of 2.5 cm/s for a distance of 50 m. Prior to the tests, the sample surfaces were prepared to have an average surface roughness of 0.04 mm. The counterpart in the ball-on-disc configuration was a 3-mm-diameter WC ball, and all tests were conducted at an ambient temperature of

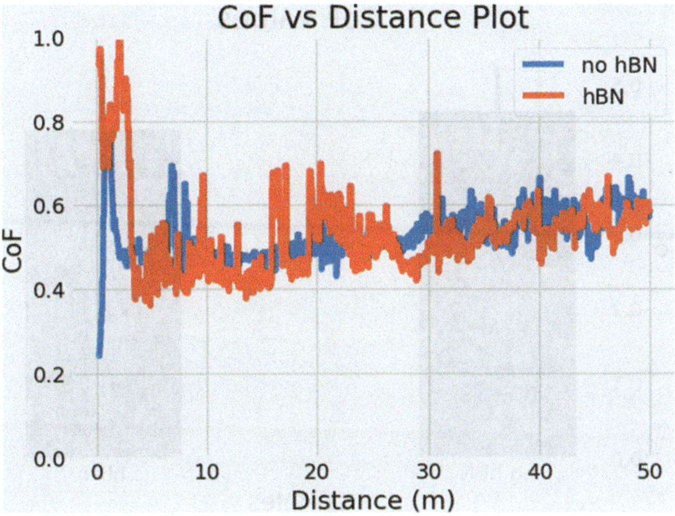

Fig. 1.2 Friction coefficient profile during the tests

25 °C and a relative humidity of 35%. Surface roughness values were measured using a Mitutoyo SJ-400 surface profilometer. A set of samples was tested with no lubrication. Another set was tested with hBN lapped with polishing cloth. A total of 3 g hBN was utilized for each sample. The hBN used in the experiments was obtained from the BORTEK Company. The results were analyzed using Python, and tribometer test data was filtered using the 500-rolling average method from the Python Pandas library [46].

According to the test results, friction coefficients with respect to the wear distance are plotted in Fig. 1.2. On the other hand, Fig. 1.3 shows the average friction coefficients for the samples. It was observed that the average coefficient of friction on the hBN-lapped surface is slightly lower than on the virgin surface. The reduction in the friction coefficient is 5.7%. Although this value is quite low, it will enable higher working conditions and lower wear. It is also important to note that there is an increase in chatter on the samples coated with hBN. This stemmed from the application procedure of the solid lubricant. The low surface roughness created areas with less hBN, and when the counterpart reached those areas, friction increased. In order to improve this behavior, laser surface texturing could be used.

1.6 Conclusion

Solid lubricants are an essential part of aircraft and aerospace vehicles. Their unique ability to perform in extreme environments where traditional lubricants fail makes them indispensable for a range of applications. Some of these applications are rocket engines, couplings, bearings, gimbals, gyroscopes, and other aerospace

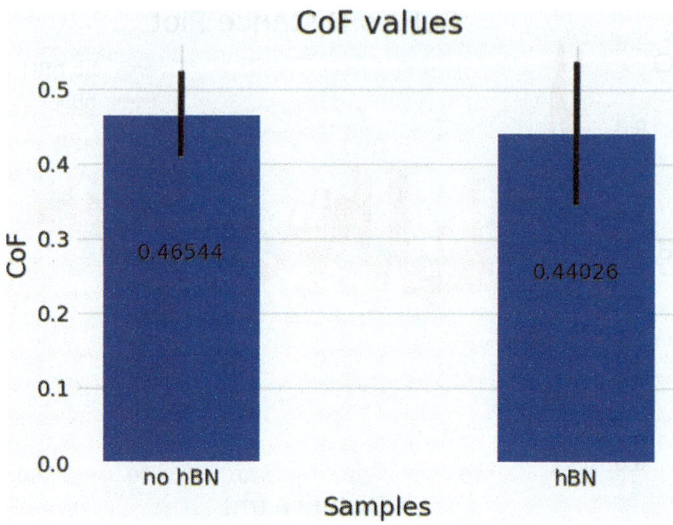

Fig. 1.3 Average friction coefficients for the samples

instrumentation, bearings for thrust and vector control, seals for valves and pumps, and space suit articulation points and joints. The solid lubricants graphene, MoS_2, and hBN are excellent examples of solid lubricants. These lamellar solid lubricants can be applied to surfaces with burnishing, bonding, or sputtering. Of these methods, the effect of burnishing hBN on AISI 4140 was investigated, and it was found that the friction coefficient of the treated component was reduced by 5.7%. This shows that applying solid lubrication to the frictional components of aircraft and aerospace vehicles improves tribological properties. Under harsh conditions, solid lubrication is a useful tool in designers' belt to prevent wear.

References

1. Holmberg K, Erdemir A. Influence of tribology on global energy consumption, costs and emissions. Friction. 2017;5(3):263–84.
2. Lince JR. Effective application of solid lubricants in spacecraft mechanisms. Lubricants. 2020;8(7):74.
3. Scharf TW, Prasad SV. Solid lubricants: a review. J Mater Sci. 2013;48(2):511–31.
4. Miyoshi K. Solid lubricants and coatings for extreme environments: state-of-the-art survey. 2007 [cited 2017 Mar 9]. Available from: https://ntrs.nasa.gov/search.jsp?R=20070010580
5. Reeves CJ, Menezes PL, Lovell MR, Jen TC. Tribology of solid lubricants. In: Menezes PL, Nosonovsky M, Ingole SP, Kailas SV, Lovell MR, editors. Tribology for scientists and engineers: from basics to advanced concepts [Internet]. New York: Springer; 2013. p. 447–94. [cited 2024 Jan 17]. Available from: https://doi.org/10.1007/978-1-4614-1945-7_13.
6. Donnet C, Erdemir A. Historical developments and new trends in tribological and solidlubricant coatings [Internet]. [cited 2011 Nov 13]. Available from: http://www.sciencedirect.com/science/article/pii/S025789720301154X

7. Spear JC, Ewers BW, Batteas JD. 2D-nanomaterials for controlling friction and wear at interfaces. Nano Today. 2015;10(3):301–14.
8. Xu M, Liang T, Shi M, Chen H. Graphene-like two-dimensional materials. Chem Rev. 2013;113(5):3766–98.
9. Ji J, Dong S, Zhang A, Zhang Q. Low-frequency interlayer vibration modes in two-dimensional layered materials. Physica E. 2016;80:130–41.
10. Ogawa S, Fukushima S, Shimatani M. Hexagonal boron nitride for photonic device applications: a review. Materials. 2023;16(5):2005.
11. Ouyang JH, Li YF, Zhang YZ, Wang YM, Wang YJ. High-temperature solid lubricants and self-lubricating composites: a critical review. Lubricants. 2022;10(8):177.
12. Lansdown AR. Molybdenum Disulphide lubrication. Amsterdam: Elsevier; 1999. p. 407.
13. Ay N, Ay GM, Göncü Y. Environmentally friendly material: hexagonal boron nitride. J Boron. 2016;1(2):66–73.
14. Gong Y, Xu ZQ, Li D, Zhang J, Aharonovich I, Zhang Y. Two-dimensional hexagonal boron nitride for building next-generation energy-efficient devices. ACS Energy Lett. 2021;6(3):985–96.
15. Marques A, Paipa Suarez M, Falco Sales W, Rocha Machado Á. Turning of Inconel 718 with whisker-reinforced ceramic tools applying vegetable-based cutting fluid mixed with solid lubricants by MQL. J Mater Process Technol. 2019;266:530–43.
16. Johnston RRM, Moore AJW. The burnishing of molybdenum disulphide on to metal surfaces. Wear. 1964;7(6):498–512.
17. Morimoto T. Effect of molybdenum disulphide upon the friction and wear in ceramic–steel pair. Tribol Int. 1997;30(12):871–9.
18. Conley P, Packard D, Purdy W. Space vehicle mechanisms: elements of successful design [Internet]. New York: Wiley; 1998 [cited 2017 Mar 2]. Available from: http://books.google.com/books?id=l5ZTAAAAMAAJ
19. CRC Press [Internet]. 2006 [cited 2017 Mar 2]. Handbook of lubrication and tribology: volume I application and maintenance, 2nd ed. Available from: https://www.crcpress.com/Handbook-of-Lubrication-and-Tribology-Volume-I-Application-and-Maintenance/Totten/p/book/9780849320958
20. E-25 General Standards for Aerospace and Propulsion Systems. Lubricant Application, Solid Film, Heat Cured, Corrosion Inhibiting [Internet]. SAE International; [cited 2024 Jan 23]. Available from: https://www.sae.org/content/as5528f
21. Clauss FJ. Solid lubricants and self-lubricating solids. New York: Academic Press; 1972. p. 260.
22. Chen JM, Ye YP, Dang HX. Developing situation and application of bonded solid lubricant films. Tribology. 1994;14(2):180–9.
23. Liang J. Bonded solid lubrication coatings, process, and applications. In: Wang QJ, Chung YW, editors. Encyclopedia of tribology [Internet]. Boston: Springer US; 2013. p. 242–7. [cited 2024 Jan 25]. Available from: https://doi.org/10.1007/978-0-387-92897-5_1241.
24. Zhang D, Gan M, Jia Y, Wan H, Chen L, Zhou H, Chen J. Synthesis of CdS nanoparticles and the effect of their addition on the tribological behavior of bonded PTFE solid lubricant coating. Tribology. 2016;36(5):629–35.
25. Yang K, Yan H, Zhang Y, Liu R. Properties of boron nitride/hyperbranched polysiloxane-polyimide bonded solid lubricant coating. Tribology. 2021;42(5):1053–62.
26. Yang R-J, Ye Y-P, Wan H-Q, Zhou H-D, Chen J-M. Tribological properties of polyimide bonded MoS2 solid lubricant coatings in different oils. Tribology. 2011;31(3):278–82.
27. Spalvins T. Lubrication with sputtered MoS2 films. E Trans. 1971;14(4):267–74.
28. Hilton MR, Jayaram G, Marks LD. Microstructure of cosputter-deposited metal- and oxide-MoS2 solid lubricant thin films. J Mater Res. 1998;13(4):1022–32.
29. Scharf TW, Kotula PG, Prasad SV. Friction and wear mechanisms in MoS2/Sb2O3/Au nanocomposite coatings. Acta Mater. 2010;58(12):4100–9.
30. Mittal D, Singh D, Sharma SK, Mittal D, Singh D, Sharma SK. Thermal characteristics and tribological performances of solid lubricants: a mini review. In: Advances in rheology of

materials [Internet]. IntechOpen; 2023 [cited 2024 Jan 25]. Available from: https://www.intechopen.com/chapters/86047
31. Allam IM. Solid lubricants for applications at elevated temperatures. J Mater Sci. 1991;26(15):3977–84.
32. Gong T, Yao P, Zuo X, Zhang Z, Xiao Y, Zhao L, et al. Influence of WC carbide particle size on the microstructure and abrasive wear behavior of WC–10Co–4Cr coatings for aircraft landing gear. Wear. 2016;362–363:135–45.
33. Gautier G, Faga MG, Tebaldo V. Impact Wear resistance of nanocomposite coatings for aircraft components. Key Eng Mater. 2019;813:387–92.
34. CS-29 Initial Issue – decision on certification specifications for large rotorcraft [Internet]. The Executive Director of the European Aviation Safety Agency; 2003. Available from: https://www.easa.europa.eu/en/document-library/certification-specifications/group/cs-29-large-rotorcraft#cs-29-large-rotorcraft
35. Posmyk A, Myalski J. Coatings protecting against aviation piston engine seizure. Wear. 2023;518–519:204627.
36. Hilton MR, Fleischauer PD. Applications of solid lubricant films in spacecraft. Surf Coat Technol. 1992;54–55:435–41.
37. Bhat A, Budholiya S, Raj SA, Sultan MTH, Hui D, Shah AUM, et al. Review on nanocomposites based on aerospace applications. Nanotechnol Rev. 2021;10(1):237–53.
38. Kumar JA, Prakash P, Krithiga T, Amarnath DJ, Premkumar J, Rajamohan N, et al. Methods of synthesis, characteristics, and environmental applications of MXene: a comprehensive review. Chemosphere. 2022;286:131607.
39. Serles P, Hamidinejad M, Demingos PG, Ma L, Barri N, Taylor H, et al. Friction of Ti3C2Tx MXenes. Nano Lett. 2022;22(8):3356–63.
40. Munkhbat B, Yankovich AB, Baranov DG, Verre R, Olsson E, Shegai TO. Transition metal dichalcogenide metamaterials with atomic precision. Nat Commun. 2020;11(1):4604.
41. Xiao Y, Xiong C, Chen MM, Wang S, Fu L, Zhang X. Structure modulation of two-dimensional transition metal chalcogenides: recent advances in methodology, mechanism and applications. Chem Soc Rev. 2023;52(4):1215–72.
42. Vazquez-Martinez JM, Sol Illana ID, Ukar Arrien E, Batista M, Salguero J. Chapter 18 – laser surface texturing as a finishing process for aerospace alloys. In: Gupta K, Pramanik A, editors. Advanced machining and finishing [internet]. Elsevier; 2021. p. 643–66. [cited 2024 Feb 22]. (Handbooks in Advanced Manufacturing). Available from: https://www.sciencedirect.com/science/article/pii/B9780128174524000105.
43. Spranger F, Hilgenberg K. Laser implantation: an innovative technique for surface texturing. PhotonicsViews. 2019;16(1):38–41.
44. Yu M, Liu M, Fu S. Slippery antifouling polysiloxane–polyurea surfaces with matrix self-healing and lubricant self-replenishing. ACS Appl Mater Interfaces. 2021;13(27):32149–60.
45. Yu M, Liu M, Zhang L, Li M, Hou Y, Wang D, et al. Liquid-repellent and self-repairing lubricant-grafted surfaces constructed by thiol-ene click chemistry using activated hollow silica as the lubricant reservoir. J Colloid Interface Sci. 2021;586:279–91.
46. The Pandas Development Team. Pandas-dev/Pandas: Pandas [Internet]. Zenodo; 2024. [cited 2024 Feb 22]. Available from: https://zenodo.org/doi/10.5281/zenodo.3509134

Chapter 2
A Numerical Study on the Flexural Behavior and Failure Mechanisms of Fiber Metal Laminates

Ibrahim H. Abuzayed, Nanda Wirawan, and Jose L. Curiel-Sosa

2.1 Introduction

The concept of bonding laminates was initially introduced as a solution to reduce fatigue crack growth in structures. Researchers found this to be an interesting idea for developing new materials that could be alternatives to conventional metallic aerospace materials, where fatigue is a critical failure mode. Researchers at the Delft University of Technology studied crack growth in aluminum sheets reinforced with fibers, where they found that the loads at the cracked metallic layers were transferred via the adhesive layer to the fiber layers, resulting in a significant reduction in crack growth [1].

Fiber metal laminates (FMLs) were recognized as hybrid composite materials that integrate metal and fiber-reinforced polymer (FRP) layers. They gained a high reputation in the aerospace industry following the registration of the patent in 1981 [2]. The first commercial FML was aramid-reinforced aluminum laminates (ARALL), which were used in the wing panels of the Fokker F-27 [2]. This was followed by the development of other FMLs such as glass laminate aluminum-reinforced epoxy (GLARE), which has excellent impact properties. This material was implemented in the cargo floor of the Boeing 777 and in parts of the fuselage of the Airbus 340 [2].

It has been challenging to understand the mechanical response of FMLs since they contain layers of materials with different mechanical behaviors. Several researchers have investigated the tensile properties of FMLs. In early studies [3, 4], researchers conducted various experiments to find the blunt notch strength of FMLs considering different notch shapes and sizes. Moreover, the literature has some numerical approaches for tensile tests of FMLs as seen in previous works [4–6]. The

I. H. Abuzayed (✉) · N. Wirawan · J. L. Curiel-Sosa
Department of Mechanical Engineering, University of Sheffield, Sheffield, UK
e-mail: ihabuzayed1@sheffield.ac.uk

S. Gürgen (ed.), *Failure in Aircraft Materials*, Sustainable Aviation,
https://doi.org/10.1007/978-3-031-65850-1_2

bending behavior of FMLs received relatively less attention from researchers. The bending behavior of carbon fiber-reinforced aluminum laminates (CARALL) and GLARE was experimentally investigated through a three-point bending test by Ostapiuk et al. [7]. Furthermore, Liu et al. [8] studied the inter-laminar behavior of GLARE using a short beam three-point bending test. In a recent study, Yao et al. [9] investigated the three-point bending of CARALL both numerically and experimentally.

This chapter presents a comprehensive damage model for predicting the failure modes in FML's constituents. The model predicts fiber failures based on Hashin's failure criteria and matrix cracking based on Puck's 3D fracture plane theory. Furthermore, a search algorithm was implemented to find the fracture angle in a computationally efficient way. The mathematical formulation of the damage initiation and propagation was given as a user-defined subroutine (VUMAT) and was used along with ductile and cohesive models in ABAQUS to simulate the failure modes of CARALL in the three-point bending test. This approach proposes an original combination of techniques, criteria, models, and a search algorithm.

2.2 Progressive Damage of FMLs

The current damage model integrates several criteria, an interface model, and a damage evolution law aimed at simulating the mechanical behavior and failure mechanisms occurring in FMLs under various types of mechanical tests. However, the focus of this study is on the bending behavior of FMLs. The current composite damage model was developed based on continuum damage mechanics, assuming transversely isotropic material behavior, where the onset of damage in fibers and matrix was assessed based on the 3D failure criteria of Hashin and Puck. Damage propagation was modeled using the equivalent displacement evolution law proposed by Lapczyk et al. [6]. The implementation of this formulation is intended to cover some of the limitations of the plane stress damage models. Moreover, the current model incorporates ductile and cohesive surface models, where the full description of these models can be found in a previous study [5].

2.2.1 3D Damage Model of Composites

2.2.1.1 Fiber Failure Criteria

The damage initiation in fibers was predicted both in tension and compression according to Hashin's criteria [10]. The criteria are expressed based on the strength along the fiber direction and the in-plane and out-of-plane shear stresses, as shown in Eqs. 2.1 and 2.2.

Fiber tension ($\hat{\sigma}_{11} \geq 0$):

$$F_{ft} = \left(\frac{\hat{\sigma}_{11}}{X^T}\right)^2 + \left(\frac{\hat{\sigma}_{12}}{S^L}\right)^2 + \left(\frac{\hat{\sigma}_{13}}{S^L}\right)^2 = 1 \tag{2.1}$$

Fiber compression ($\hat{\sigma}_{11} < 0$):

$$F_{fc} = \left(\frac{\hat{\sigma}_{11}}{X^C}\right)^2 = 1 \tag{2.2}$$

2.2.1.2 Matrix Failure Criteria

Matrix failure initiation was assessed based on Puck's inter-fiber fracture (IFF) criterion. The criterion depends on the 3D stresses acting on an action plane, where the basics of Puck's formulations are based on the brittle behavior of composites [11]. This enables the IFF to compute the stress at which the fracture is initiated and predict the fracture angle, which gives an indication of the crack direction. The IFF formulations consider the analysis of stress on the fracture plane; thus, the transformation matrix shown in Eq. 2.3 is used to find stresses (σ_n, τ_{nt}, τ_{nl}), and out-of-normal stresses (X_1, X_2, X_3). The relationship between these stresses is illustrated in Fig. 2.1.

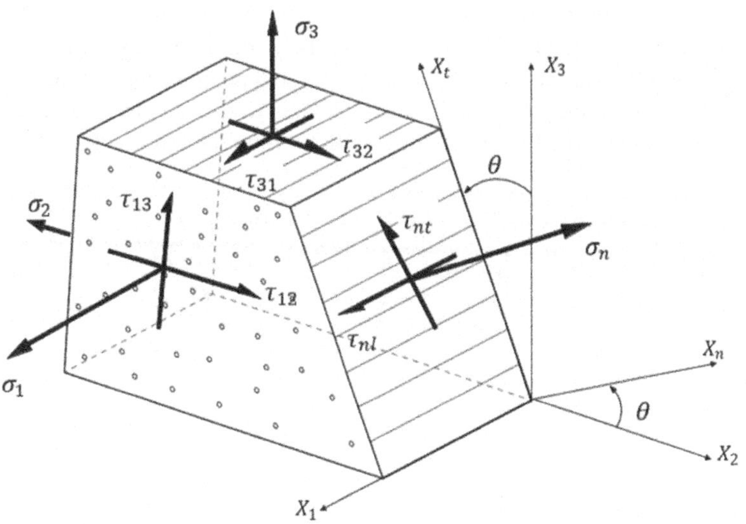

Fig. 2.1 Definition of stresses (σ_n, τ_{nt}, τ_{n1}) on fiber parallel plane

$$\begin{bmatrix} \sigma_n(\theta) \\ \tau_{nt}(\theta) \\ \tau_{n1}(\theta) \end{bmatrix} = \begin{bmatrix} \cos^2(\theta) & \sin^2(\theta) & 2\cos(\theta)\sin(\theta) & 0 & 0 \\ -\cos(\theta)\sin(\theta) & \cos(\theta)\sin(\theta) & \cos^2(\theta)-\sin^2(\theta) & 0 & 0 \\ 0 & 0 & 0 & \sin(\theta) & \cos(\theta) \end{bmatrix} \begin{bmatrix} \sigma_{22} \\ \sigma_{33} \\ \sigma_{23} \\ \sigma_{13} \\ \sigma_{12} \end{bmatrix} \tag{2.3}$$

Puck defined three resistances $\left(R_{\parallel\perp}^{At}, R_{\perp\perp}^{A}, R_{\perp}^{A}\right)$ at the action plane, where the subscripts $\perp$ and $\parallel$ define the directions perpendicular or parallel to the fibers' direction, which are unusually expressed as 1 along the fiber and 2 across the fiber. ($R_{\parallel\perp}^{At}$) and ($R_{\perp}^{A}$) indicate the resistance of the action plane against fracture resulting from longitudinal shear, τ_{21}, and tensile, σ_{22}, stresses, respectively [12]. These stresses can be expressed as ($R_{\parallel\perp}^{At}$) = Y_t and $\left(R_{\perp}^{A}\right) = S_L$ at the fracture plane. The third fracture resistance, $R_{\perp\perp}^{A}$, is due to transverse shear stress, τ_{21}. However, it does not equal the shear stress, $\tau_{\perp\perp}$, as the fracture of brittle materials due to an applied stress, $\tau_{\perp\perp}$, does not occur at the fracture plane [13].

$$R_{\perp\perp}^{A} = \frac{R_{\perp}^{c}}{2\left(1 + p_{\perp\perp}^{c}\right)} \tag{2.4}$$

$R_{\perp\perp}^{A}$ is found by using Eq. 2.4, where $R_{\perp}{}^{c}$ is the transverse compressive strength (Y_c), and $p_{\perp\perp}^{c}$ is the inclination parameter. The damage exposure factors, which define the initiation of damage in the IFF criterion, are calculated by using Eqs. 2.5 and 2.6 for tensile and compressive loads, respectively. The values of the inclination parameters ($p_{\perp\perp}^{t,c}$, $p_{\perp\parallel}^{t,c}$) are defined in Table 2.1. The inclination parameters at angle ψ are calculated by using Eq. 2.7 with $cos^2\psi$ and $sin^2\psi$ expressed in terms of the fracture plane shear stresses in Eqs. 2.8 and 2.9. The fracture plane's orientation (θ_{fp}) varies in the range of -90° to 90°, while the maximum value of $f_{E,IFF+}$ or $f_{E,IFF-}$ defines the fracture angle (θ_{fp}). Puck suggested using small increments ($\Delta\theta = 1$), which requires 180 computations for a single Gaussian point. The next section will discuss various numerical methods for calculating the fracture angle (θ_{fp}).

For $\sigma_n \geq 0$,

$$f_{E,IFF+}(\theta) = \sqrt{\left[\left(\frac{1}{R_{\perp}^{At}} - \frac{p_{\perp\psi}^{t}}{R_{\perp\psi}^{A}}\right)\sigma_n(\theta)\right]^2 + \left(\frac{\tau_{nt}(\theta)}{R_{\perp\perp}^{A}}\right)^2 + \left(\frac{\tau_{n1}(\theta)}{R_{\parallel\perp}^{A}}\right)^2} + \frac{p_{\perp\psi}^{t}}{R_{\perp\psi}^{A}}\sigma_n(\theta) \tag{2.5}$$

Table 2.1 Inclination parameters for carbon fiber laminates [13]

$p_{\perp\parallel}^{t}$	$p_{\perp\parallel}^{c}$	$p_{\perp\perp}^{t}$	$p_{\perp\perp}^{c}$
0.35	0.30	0.25	0.30

For $\sigma_n < 0$,

$$f_{E,IFF-}(\theta)=\sqrt{\left(\frac{p_{\perp\psi}^{c}}{R_{\perp\psi}^{A}}\sigma_n(\theta)\right)^2+\left(\frac{\tau_{nt}(\theta)}{R_{\perp\perp}^{A}}\right)^2+\left(\frac{\tau_{n1}(\theta)}{R_{\parallel\perp}^{A}}\right)^2}+\frac{p_{\perp\psi}^{c}}{R_{\perp\psi}^{A}}\sigma_n(\theta) \tag{2.6}$$

$$\frac{p_{\perp\psi}^{t,c}}{R_{\perp\psi}^{A}}=\frac{p_{\perp\perp}^{t,c}}{R_{\perp\perp}^{A}}cos^2\psi+\frac{p_{\perp\parallel}^{t,c}}{R_{\perp\parallel}^{A}}sin^2\psi \tag{2.7}$$

$$cos^2\psi=\frac{\tau_{nt}}{\tau_{nt}^2+\tau_{nl}^2} \tag{2.8}$$

$$sin^2\psi=\frac{\tau_{nl}}{\tau_{nt}^2+\tau_{nl}^2} \tag{2.9}$$

2.2.1.3 Fracture Plane Angle Search Algorithm

Puck utilized the stepwise search method (SSM) to obtain the fracture plane angle orientation. In this method, the failure exposure value is calculated for each inclination angle, θ from $-90^\circ \leq \theta < 90^\circ$ with a 1° increment. The results are then compared to determine the largest failure exposure value and its corresponding inclination angle, θ. This will guarantee that the fracture plane angle orientation will have the inclination angle, θ, containing the largest failure exposure value. However, the SSM requires a significant amount of time to complete since it has to calculate 180 failure exposure values within $-90^\circ \leq \theta < 90^\circ$ range. Therefore, several fast fracture plane angle orientation search algorithms have been developed to reduce the number of required failure exposure calculations without sacrificing the accuracy of the result. An example of failure exposure—inclination angle calculation using the SSM algorithm is plotted in Fig. 2.2.

Weigand et al. [14] proposed the extended golden section search (EGSS) to obtain the fracture plane angle orientation. In the EGSS algorithm, Weigand combines the golden section search (GSS) algorithm with a successive parabolic interpolation method. In this method, the GSS algorithm calculates the necessary failure exposure value until the last two failure exposure values have corresponding inclination angles within the predetermined value. The successive parabolic interpolation method is then used to determine the fracture plane angle orientation by substituting the last two failure exposure values and their corresponding inclination angles into Eq. 2.10. The EGSS dramatically reduces the total number of failure exposure value calculations required to obtain the fracture plane angle orientation, thus reducing computation time. Figure 2.3 shows the plot of failure exposure—inclination angle plot of the EGSS algorithm.

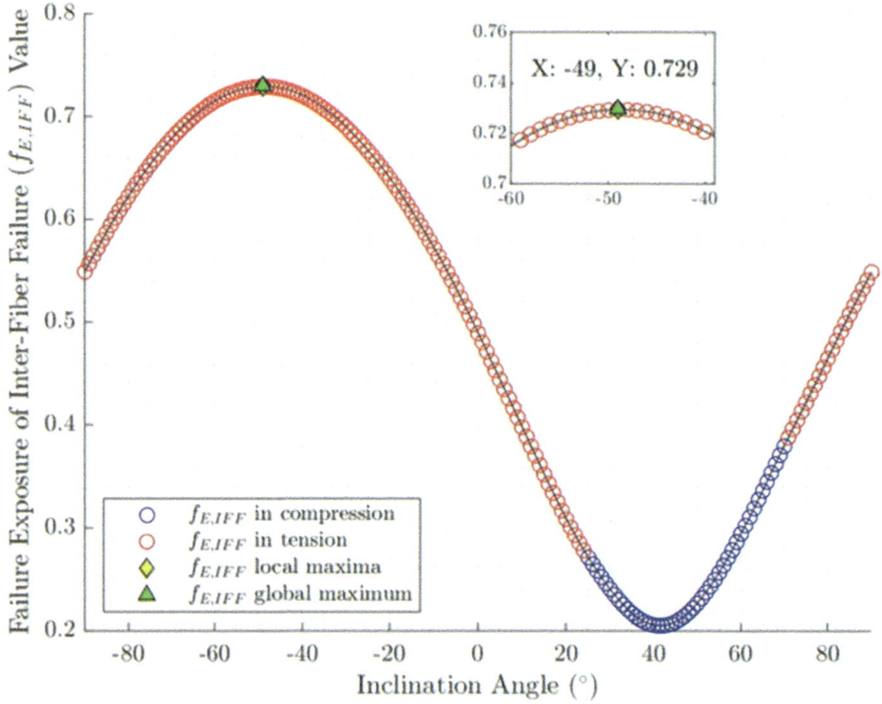

Fig. 2.2 Failure exposure—inclination angle plot using SSM algorithm

$$\theta_{fp} = \theta_2 - \frac{1}{2} \frac{(\theta_2 - \theta_1)^2 \left(f_{E,IFF(2)} - f_{E,IFF(3)}\right) - (\theta_2 - \theta_3)^2 \left(f_{E,IFF(2)} - f_{E,IFF(1)}\right)}{(\theta_2 - \theta_1)\left(f_{E,IFF(2)} - f_{E,IFF(3)}\right) - (\theta_2 - \theta_3)\left(f_{E,IFF(2)} - f_{E,IFF(1)}\right)} \tag{2.10}$$

Schirmaier et al. [15] found that most of the time, the EGSS algorithm is unable to correctly identify the global maximum value of failure exposure when there are two or more local maximum failure exposure values (Fig. 2.4). They suggest a more reliable fast fracture plane angle orientation search algorithm called the selective range golden section search (SRGSS) algorithm.

In the SRGSS algorithm, a selective range (SR) method is employed before applying the GSS algorithm. The SR method calculates all failure exposure values from $-90^\circ \leq \theta < 90^\circ$ with 10° intervals. Each failure exposure value is compared with the neighboring failure exposure values. If the current failure exposure value is larger than its neighboring failure exposure values, then those intervals are selected as the SR interval. The GSS algorithm is applied to all of the SR intervals to obtain the local maximum failure exposure value on each SR interval. In the last step, all of the local maximum failure exposure values are compared to obtain the global maximum failure exposure value. Figure 2.5 shows the SRGSS algorithm, correctly obtaining the fracture plane angle orientation.

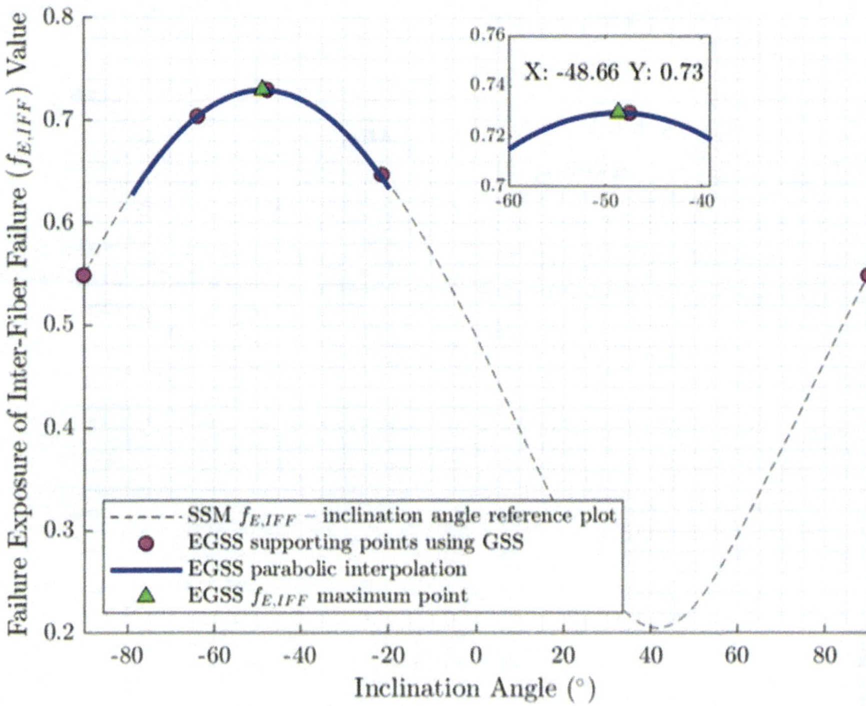

Fig. 2.3 Failure exposure—fracture angle plot using EGSS algorithm

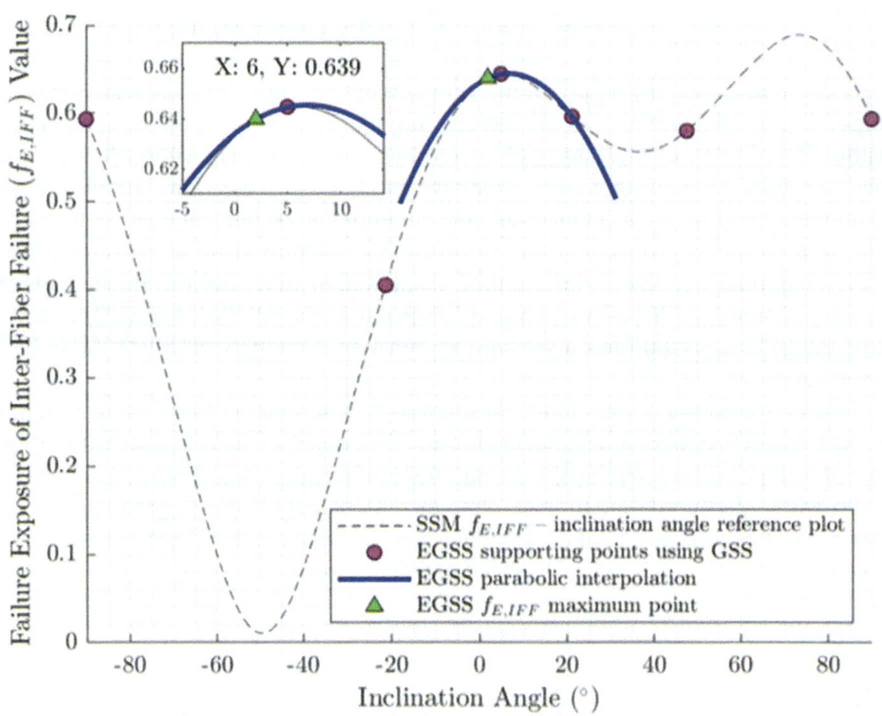

Fig. 2.4 Failure exposure—fracture angle plot with two local maximum using EGSS algorithm

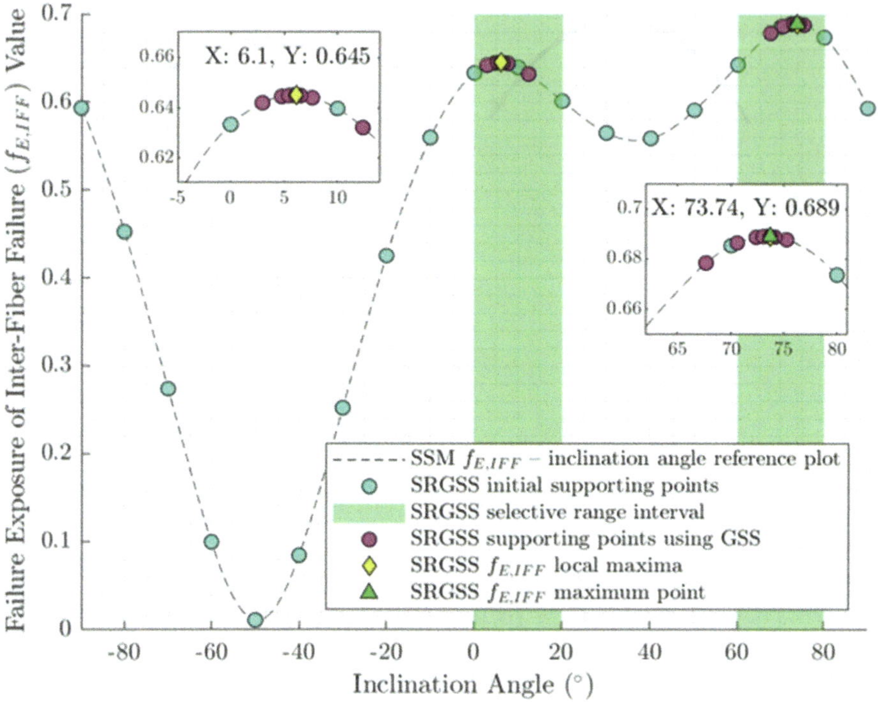

Fig. 2.5 Failure exposure—fracture angle plot with two local maximum using SRGSS algorithm

Rezasefat et al. [16] proposed the simple parabolic interpolation search (SPIS) algorithm as an alternative to the fast fracture plane angle orientation search algorithm. The SPIS uses the same SR method as the SRGSS algorithm; however, it replaces the GSS algorithm with successive parabolic interpolation. The successive parabolic interpolation is applied only once within the SR interval that has the largest average failure exposure value to obtain the fracture plane angle orientation. This combination reduces the required computation time, but the fracture plane angle orientation will be less accurate compared to the SRGSS. Figure 2.6 shows that the SPIS algorithm has a similar capability to accurately calculate the failure exposure.

In order to simulate the failure of composite materials by using Puck's failure criterion in ABAQUS finite element analysis (FEA) software, a separate subroutine was developed. The failure exposure and fracture plane angle orientation are calculated within the subroutine for each time increment and Gauss/integration point by using the material and strain input data from ABAQUS. The SPIS algorithm is implemented in the subroutine to reduce computation time.

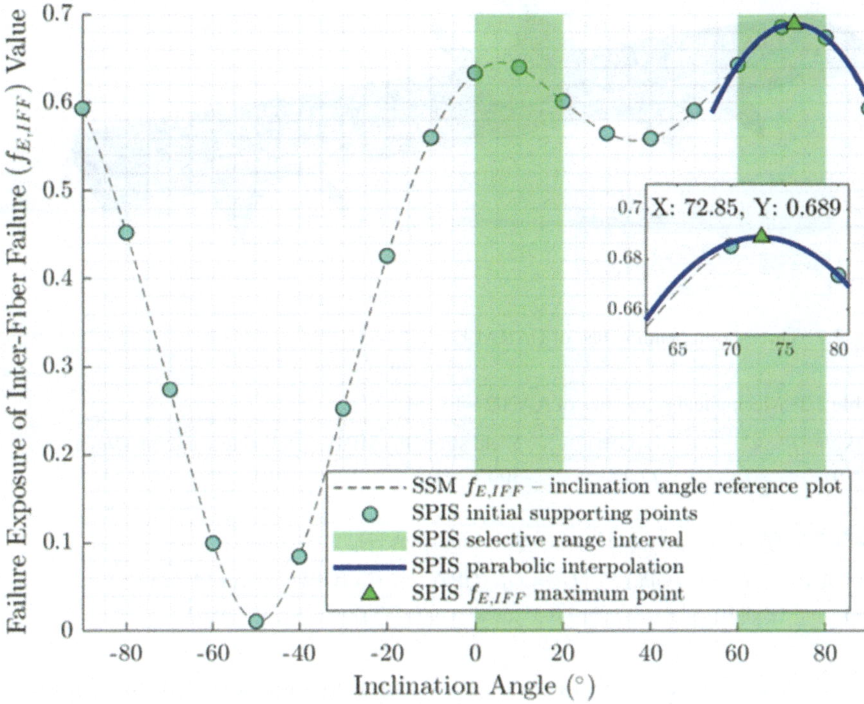

Fig. 2.6 Failure exposure—fracture angle plot with two local maxima using SPIS algorithm

2.3 Three-Point Bending Test of CARALL

The three-point bending is a mechanical test used for characterizing the flexural properties of materials, i.e., flexural modulus and strength. According to the ASTM standard of the three-point bending of polymer matrix composite materials (D7264), the test is performed on a rectangular cross-sectional specimen resting on two supporting points, while the load is applied at the center of the specimen as illustrated in Fig. 2.7. In this investigation, a CARALL specimen consisting of three AA2024-T3 and two carbon fiber unidirectional laminates is considered. The thickness of each aluminum sheet is 0.5 mm, and each carbon fiber laminate has a $[0^{\circ},90^{\circ}]_s$ laminate with a thickness of 0.5 mm. The total length of the specimen is 80 mm, whereas the width and total thickness are 10 mm and 2.5 mm, respectively. The loading and support noses have a diameter of 10 mm, and the distance between the two supports is 45 mm.

In the FE model, all degrees of freedom are fixed for the supports, and a vertical displacement is applied at the top nose. A general contact behavior was defined

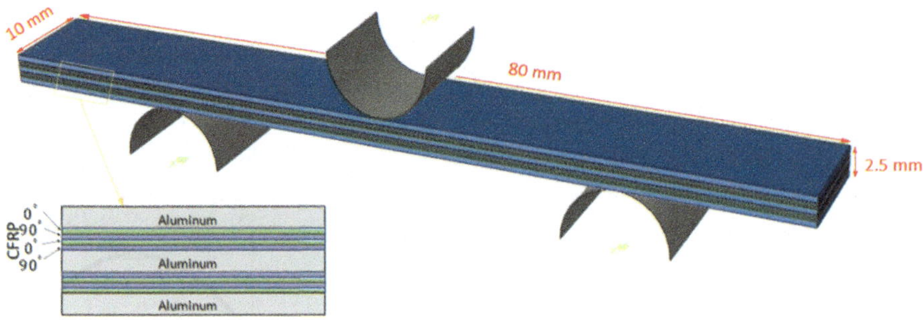

Fig. 2.7 Three-point bending test of CARALL

Table 2.2 Mechanical properties of AA2024-T3 [6]

ρ (kg/m^3)	ν	E (GPa)	Yield strength (MPa)	Fracture energy (J/m^2)
2770	0.33	72.2	300	10.2

Table 2.3 Elastic properties of carbon fiber laminates [9]

E_1 (GPa)	$E_2 = E_3$ (GPa)	$G_{12} = G_{13}$ (GPa)	G_{23} (GPa)	$\nu 12 = \nu 13$
108	8.0	4.0	3.0	0.32

between the noses and specimen with a friction coefficient of 1.0. In addition, a cohesive interaction was defined at the interface between the carbon fiber laminates and aluminum sheets. All layers were constructed with the 8-node linear brick elements with reduced integration (C3D8R). The mesh size was refined to 0.25 mm at the middle area, where failure is expected.

2.3.1 *Material Properties of CARALL*

Elasto-plastic behavior was assumed for the aluminum sheets. The material properties of AA2024-T3 are given in Table 2.2. Tables 2.3 and 2.4 give the elastic and strength properties of carbon fiber laminates, respectively. The fracture energies of carbon fiber laminates are given in Table 2.5.

2.4 Results and Discussion

The damage mechanisms observed in the three-point bending test of CARALL are fiber breakage, matrix cracking, delamination, and plastic damage of aluminum layers. The progressive damage of each failure mode is discussed in the following sections.

Table 2.4 Strength properties of carbon fiber laminates [9]

X^T (MPa)	X^C (MPa)	Y^T (MPa)	Y^C (MPa)	S^L (MPa)
2100	800	25	120	80

Table 2.5 Fracture energies of carbon fiber laminates [17]

G_{ft} (N/mm)	G_{fc} (N/mm)	G_{mt} (N/mm)	G_{mc} (N/mm)
81.5	106.3	0.2774	1.3092

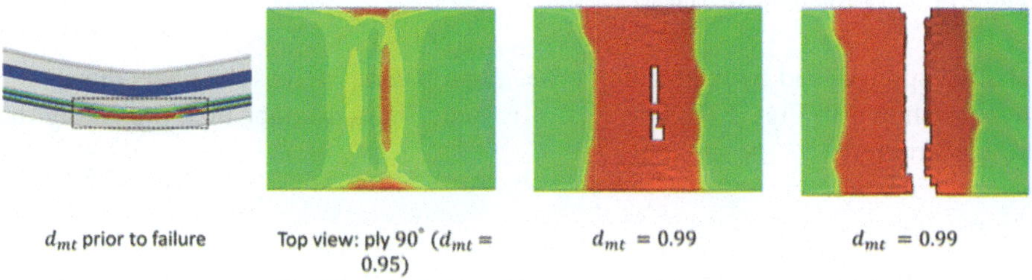

Fig. 2.8 Matrix tensile damage

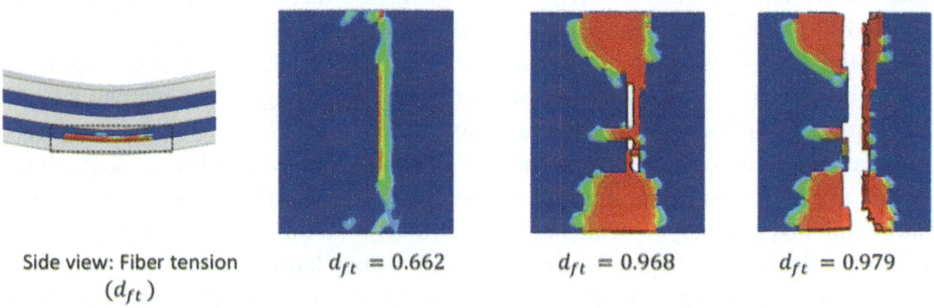

Fig. 2.9 Fiber tensile damage

2.4.1 Progressive Damage in Carbon Fiber Laminates

The progressive damage process initiates below the neutral axis in the carbon fiber laminates, where matrix tensile damage (d_{mt}) is the first failure mode during the flexural test. Matrix tensile damage is localized in the 90° layer in the middle and at the edges of the specimen; then the cracks propagate across the specimen at 0.99 of d_{mt}, as illustrated in Fig. 2.8. Indeed, the crack initiates at the center of the specimen in the 90° layer and then propagates into the adjacent 0° layer. Fiber tensile damage (d_{ft}) is mostly evident in the 0° layer; however, the damage value is far from the failure value. Nevertheless, d_{ft} increases dramatically when cracks propagate into the adjacent 90° layer, as shown in Fig. 2.9. Above the neutral axis, no evident damage is observed in the carbon fiber laminates except for fiber compression damage (d_{fc}), which begins at an early stage in the 0° layer as a result of specimen bending

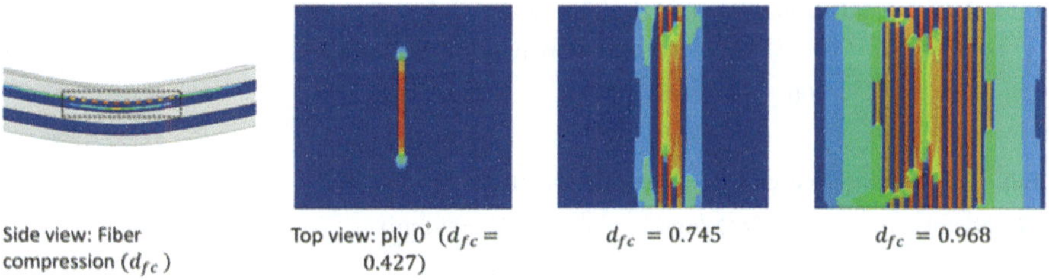

Fig. 2.10 Fiber compressive damage

and the applied load by the loading nose. d_{fc} propagates across the fibers during the test, as shown in Fig. 2.10; however, it is quite limited rather than prevailing the whole structure.

2.4.2 Progressive Damage in Aluminum Layers

The ductile damage (DUCTCRT) is plotted for the three aluminum layers of the CARALL in Fig. 2.11. The most critical ductile damage is at the bottom layer, where the damage is 0.576 before the crack initiation in the carbon fiber laminates. Once the first crack propagates in the specimen, the damage value increases to 1.0 since most of the tensile loading is concentrated in the bottom aluminum layer. Thereafter, the aluminum layer cannot withstand the stress and ultimately shows failure in a short time span. Further loading results in ductile damage in the middle aluminum layer, which is located at the neutral axis. In this layer, crack propagation is observed after the failure of the first carbon fiber laminate. The top aluminum layer experiences minor ductile damage as a result of the loading applied by the nose. The ductile damage propagates in this layer due to the bending load; however, the test is terminated before the full failure.

2.4.3 Delamination

Delamination is mostly evident at the interface of the middle aluminum layer and lower carbon fiber laminate. The contact stress damage initiation variable (CSMAXSCRT) is plotted at the aluminum layer's 90° interface prior to failure, as shown in Fig. 2.12a. The figure shows a maximum value of 0.78 at the edges of the specimen while having a very low value at the center. At the moment of crack propagation in the lower carbon fiber laminate, delamination significantly increases at the center of the specimen, as shown on the right side of Fig. 2.12b.

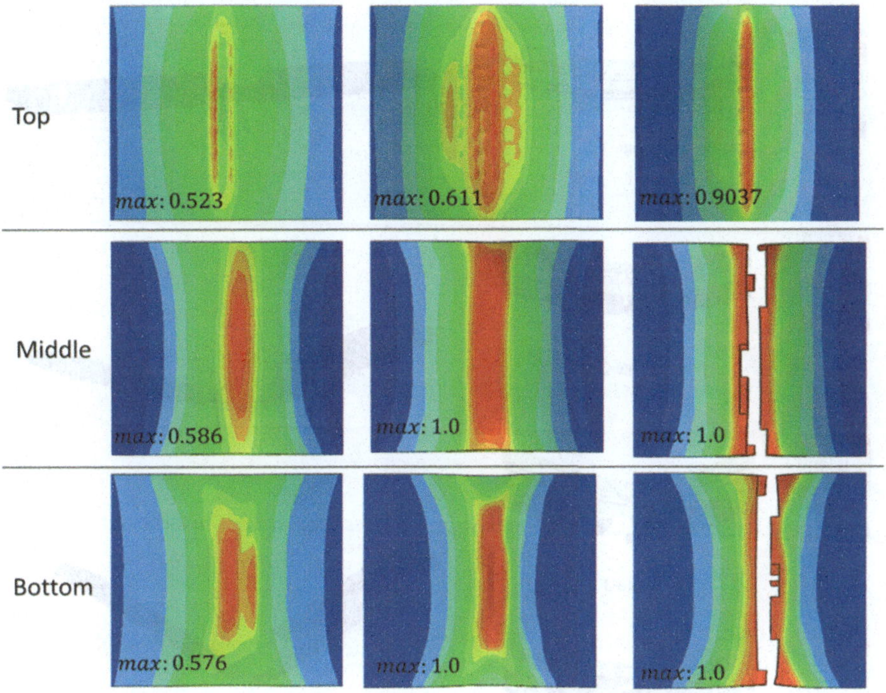

Fig. 2.11 Ductile damage at top, middle, and bottom aluminum layers

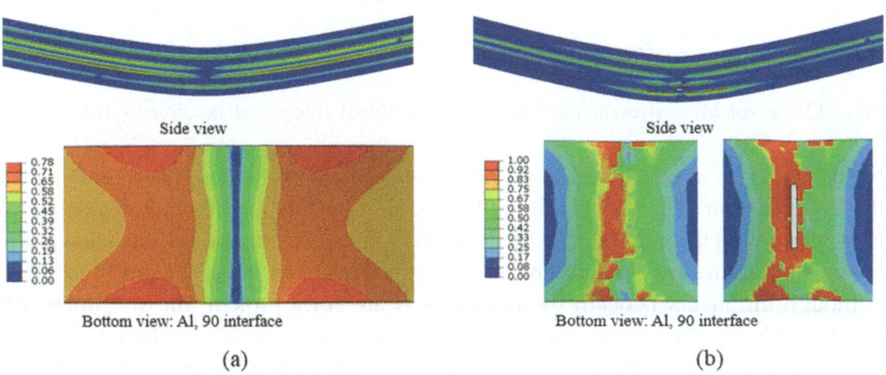

Fig. 2.12 (**a**) CSMAXSCRT before failure and (**b**) at the moment of matrix cracking

2.4.4 Failure Sequence of CARALL Under Bending Load

The overall flexural damage of CARALL, along with a detailed investigation of crack propagation, is shown in Fig. 2.13. The first failure mode observed in the fracture is matrix cracking in the 90° of the carbon fiber laminates, followed by fiber

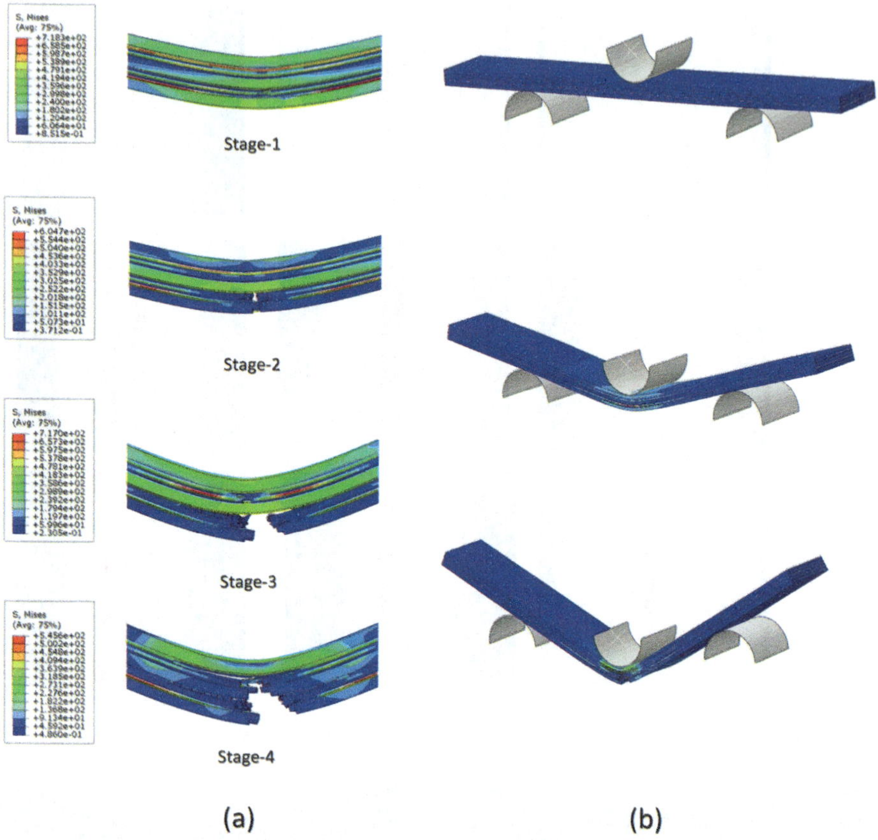

Fig. 2.13 (**a**) von Mises stress at four stages of damage and (**b**) the overall deformation in CARALL

breakage in the same layer. These failures lead to excessive delamination between the lower carbon fiber and aluminum layers. High tensile loading below the neutral axis, along with the gap induced by delamination, causes stress concentration at the lower aluminum layer, which ends up in failure as seen in stage-3 in Fig. 2.13a. Although the layers beneath the neutral axis are fully broken, the specimen still resists the load in the upper part in stage-3. The load is carried by the intact upper part of two layers of aluminum and one layer of carbon fiber laminate. Carbon fiber layers bear higher stresses due to their brittle behavior, leading to their failure first. Then, delamination develops on both sides of the carbon fiber laminates. Subsequently, aluminum layers exhibit ductile failure as illustrated in stage-4.

2.5 Conclusion

A progressive damage model is developed for composite materials using a combination of Puck's and Hashin's failure criteria. In addition, a search algorithm, which depends on parabolic interpolation, is implemented to find the fracture angle in a computationally efficient manner. This is combined with cohesive surface and ductile damage models available in ABAQUS/explicit to build a comprehensive damage model for FMLs. The model is used to study the progressive failure of CARALL under flexural load. The FEA results show that the failure of the CARALL starts in the layers below the neutral axis. Matrix cracking and fiber breakage are also observed in the carbon fiber layers. The failure process prevails over the structure with delamination and the ductile failure of the aluminum layers.

References

1. Alderliesten R. Fatigue and fracture of fibre metal laminates, vol. 236. Springer; 2017. https://doi.org/10.1007/978-3-319-56227-8.
2. Ad Vlot and Jan Willem Gunnink. Fibre metal laminates: an introduction. Springer Science & Business Media; 2001. ISBN: 978-94-010-0995-9
3. De Vries TJ. Blunt and sharp notch behaviour of glare laminates. Ph.D dissertation; 2001.
4. He W, Wang C, Wang S, Yao L, Wang L, Xie D. Characterizing and predicting the tensile mechanical behavior and failure mechanisms of notched FMLs—combined with DIC and numerical techniques. Compos Struct. 2020;254:112893. ISSN 0263-8223. https://doi.org/10.1016/j.compstruct.2020.112893. URL: https://www.sciencedirect.com/science/article/pii/S0263822320328191
5. Abuzayed IH, Wirawan N, Zhang C, Curiel-Sosa JL. A 3d damage model for simulating damage modes in fibre metal laminates. In: In 9th ECCOMAS thematic conference on the mechanical response of composites. CIMNE; 2023. https://doi.org/10.23967/c.composite.2023.011. URL: https://www.scipedia.com/public/Abuzayed_et_al_2023a.
6. Lapczyk I, Hurtado JA. Progressive damage modeling in fiber reinforced materials. Compos Part A Appl Sci Manuf. 2007;38(11):2333–41. ISSN 1359-835X
7. Ostapiuk M, Bieniaś J, Surowska B. Analysis of the bending and failure of fiber metal laminates based on glass and carbon fibers. Sci Eng Compos Mater. 2018;25(6):1095–106. https://doi.org/10.1515/secm-2017-0180.
8. Liu C, Dandan D, Li H, Yubing H, Yiwei X, Tian J, Tao G, Tao J. Interlaminar failure behavior of glare laminates under short-beam three-point-bending load. Compos B Eng. 2016;97:361–7. ISSN 1359-8368. https://doi.org/10.1016/j.compositesb.2016.05.003. URL: https://www.sciencedirect.com/science/article/pii/S135983681630573X.
9. Yao L, Mao L, Wang Y, He W, Ma Y, Yu H. Insitu damage monitoring and numerical characterization of three-point bending and incremental cycle flexural behavior of fmls. Mech Adv Mater Struct. 2023:1–19. https://doi.org/10.1080/15376494.2023.2204095.
10. Hashin Z. Fatigue failure criteria for unidirectional fiber composites. J Appl Mech. 1981;48(4):846–52. ISSN 0021-8936. https://doi.org/10.1115/1.3157744.
11. H Matthias Deuschle and Alfred Puck. Application of the puck failure theory for fibre-reinforced composites under three-dimensional stress: comparison with experimental results. J Compos Mater, 47(6–7):827–846, 2013. https://doi.org/10.1177/0021998312462158.

12. Knops M. Analysis of failure in fiber polymer laminates: the theory of Alfred Puck. Springer; 2008. https://doi.org/10.1007/978-3-540-75765-8.
13. H Matthias Deuschle and Bernd-H Kröplin. Finite element implementation of puck's failure theory for fibre-reinforced composites under three-dimensional stress. J Compos Mater, 46(19–20):2485–2513, 2012. https://doi.org/10.1177/0021998312451480.
14. Wiegand J, Petrinic N, Elliott B. An algorithm for determination of the fracture angle for the three-dimensional puck matrix failure criterion for UD composites. Compos Sci Technol. 2008;68(12):2511–7. ISSN 0266-3538. https://doi.org/10.1016/j.compscitech.2008.05.004. URL: https://www.sciencedirect.com/science/article/pii/S0266353808001851. Deformation and Fracture of Composites: Analytical, Numerical and Experimental Techniques, with regular papers
15. Schirmaier FJ, Weiland J, Kärger L, Henning F. A new efficient and reliable algorithm to determine the fracture angle for puck's 3D matrix failure criterion for UD composites. Compos Sci Technol. 2014;100:19–25. ISSN 02663538. https://doi.org/10.1016/j.compscitech.2014.05.033. URL: https://www.sciencedirect.com/science/article/pii/S0266353814001882.
16. Rezasefat M, Torres DB, Alvaro Gonzalez-Jimenez M, Giglio, and A. Manes. A fast fracture plane orientation search algorithm for puck's 3D IFF criterion for UD composites. Mater Today Comm. 2021;28:102700. ISSN 2352-4928. https://doi.org/10.1016/j.mtcomm.2021.102700. URL: https://www.sciencedirect.com/science/article/pii/S2352492821006929.
17. Reinoso J, Catalanotti G, Blázquez A, Areias P, Camanho PP, París F. A consistent anisotropic damage model for laminated fiber-reinforced composites using the 3D-version of the puck failure criterion. Int J Solids Struct. 2017;126:37–53.

Chapter 3
Fracture Characteristics of Welded Joints in Aluminum Matrix Composites

Kelvii Wei Guo

3.1 Introduction

The significant advancements achieved in metal matrix composites (MMCs) can be attributed to the long-term dedication and enthusiasm of researchers who have contributed to their development and applications. Despite being considered mature materials, there are still limitations in many cases, prompting researchers to investigate cutting-edge theories and explore various aspects with profound effects [1–8].

MMCs, as a type of composite material, exhibit enhanced properties such as conductivity, specific strength, specific capacity, thermal expansion, and electromagnetism compared to conventional materials. Moreover, with advancements in nanotechnology, nanocomposite materials have been extensively researched, holding promise for cost-effective practical applications [6, 9–18]. The exploration of nanocomposites opens up new avenues for the development of lightweight, high-performance materials with tailored properties, paving the way for innovations in various industries ranging from aerospace and automotive to electronics and energy storage.

It is evident that MMCs are typically more expensive than the traditional materials from which they are derived. However, through the utilization of innovative techniques, this issue has been effectively addressed, resulting in a significant reduction in the cost of MMCs. These innovative approaches encompass advancements in manufacturing processes, material sourcing, and composite fabrication methods, among others. By implementing such techniques, the cost-effectiveness of MMCs has been enhanced, making them more accessible and viable for a wider range of applications across various industries. However, one of the main obstacles in the widespread application of MMCs is the interface reaction between the

K. W. Guo (✉)
Department of Mechanical and Biomedical Engineering, City University of Hong Kong, Kowloon Tong, Kowloon, Hong Kong

S. Gürgen (ed.), *Failure in Aircraft Materials*, Sustainable Aviation,
https://doi.org/10.1007/978-3-031-65850-1_3

reinforcement materials and matrices. Therefore, it is imperative to systematically explore the deep theory related to interfaces for various reinforcements and matrices. This exploration must cover detailed mechanisms, particularly focusing on interface evolution under excellent wetting conditions and reinforcement distribution within matrices [19–29]. Understanding and addressing interface reactions are critical for optimizing the performance and properties of MMCs, as well as enhancing their reliability and durability in real-world applications. This systematic investigation will contribute to overcoming barriers and advancing the field of MMCs for diverse industrial applications.

Aluminum matrix composites (AMCs) are widely recognized and utilized as aircraft materials for various designed components. They offer several advantageous properties, including high rigidity, good dimensional stability, corrosion resistance, wear resistance, and specific strength [30–32]. These desirable characteristics make AMCs highly suitable for applications in aerospace engineering, where lightweight yet durable materials are essential to ensure the structural integrity and performance of aircraft components. The versatility and reliability of AMCs contribute to their extensive use across a wide range of aircraft designs and configurations.

Welding is a practical and widely preferred joining technique for processing AMCs [33–39]. The reinforcements distributed in AMCs, whether short fibers, particulates, or whiskers, impart a discontinuous characteristic to AMCs. This discontinuity poses a unique challenge for achieving high-quality welded joints in AMCs, as it influences the selection and implementation of appropriate joining approaches. Nonetheless, with careful consideration and specialized techniques, AMCs can be effectively welded to produce joints of exceptional quality, ensuring the integrity and performance of the final components in various engineering applications.

According to the above-mentioned discontinuous feature, although these methods are available for MMCs joining, there are still some critical issues that need to be urgently solved [40, 41].

1. Particle reinforcement distribution: It is well known that the properties of welded joints are directly affected by the distribution of particle reinforcement in the weld. Welded joints with uniformly distributed particle reinforcements can achieve 80% or even slightly higher strength compared to the parent substrate. If particle reinforcements are aggregated or absent (creating a "no-reinforcements-zone") in the weld, the strength of the welded joints decreases, and they may even fail to form a proper weld.
2. Interface condition between particle reinforcements and aluminum matrix: As a special feature of MMCs, various pernicious phases are generated at the interface during the high-temperature welding process. Due to the high energy input into AMCs by electron beam, laser welding, and TIG, high temperatures are distributed in the welding pool, resulting in the formation of a pernicious phase (Al_4C_3), which subsequently leads to weld failure.
3. Low efficiency and long time: For solid-state welding and diffusion welding, it often takes more than 24 hours to complete the weld, and in some cases, it may even require several days.

This work aims to provide a detailed study of the fracture characteristics of AMC welded joints at various temperatures. Additionally, the distribution of temperature in heated substrates is also discussed.

3.2 Diffusion Welding

3.2.1 Diffusion Welding Without Interlayer

Diffusion welding is a conventional technique often chosen as the primary method for joining AMCs due to its feasibility and ease of operation. However, it is well recognized that diffusion welding, being a solid-state welding process, typically requires a significant amount of time to complete. This extended duration can lead to low efficiency and productivity, posing challenges for large-scale production and time-sensitive applications. Moreover, during the diffusion welding process, the formation of brittle intermetallic compounds at the interface can occur, which may compromise the mechanical properties and integrity of the welded joints. These intermetallic compounds can reduce the ductility and toughness of the weld, making it susceptible to premature failure under load. The poor weldability of AMCs is primarily attributed to significant differences in physical and chemical properties between the aluminum matrix and the reinforcement particles. In diffusion welding, the welding temperature plays a crucial role in achieving successful welds and ensuring adequate joint strength.

Figure 3.1 shows the microstructure of welded joints at various welding temperatures using a specimen of stir-cast SiCp/A356 (reinforced with 20% volume fraction SiC particles of 12 μm mean size), with welding parameters set at a welding pressure of 5 MPa and welding time of 30 min. At a welding temperature of 470 °C, a distinct boundary is observed between two specimens. This boundary indicates weak movement of reinforcements and relatively low ability of atoms diffusion. Additionally, direct contact between reinforcements results in weak joining,

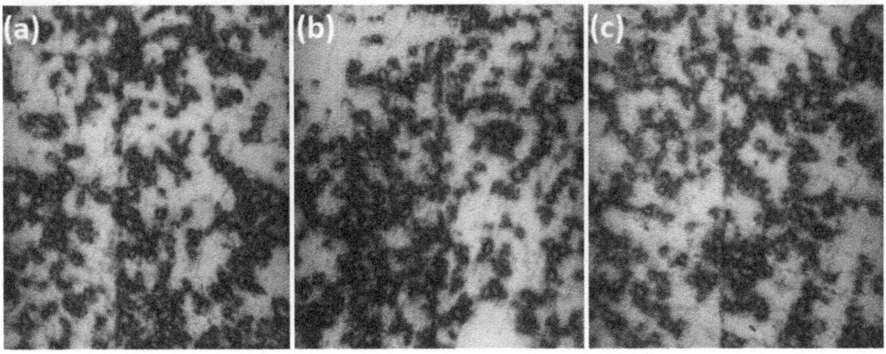

Fig. 3.1 Microstructure of joints diffusion welded at (**a**) 470 °C, (**b**) 490 °C, and (**c**) 510 °C

potentially leading to crack formation at the interface. Consequently, the welded joints exhibit low strength, measuring only 9% of the strength of the parent composites. As the welding temperature increases to 490 °C, the bonding boundary in the weld gradually disappears. This is attributed to suitable deformation of the joint, which enhances the distribution of reinforcements in the bonding boundary area. Consequently, more reinforcements aggregate at the bonding boundary, and some can be found near the joint's boundary. However, the distribution of reinforcements remains nonuniform, and particles mix in the bonding boundary area, making it difficult to discern their origins. Despite these challenges, the strength of the welded joint increases to approximately 21% of the strength of the parent composites. Upon further increasing the welding temperature to 510 °C, the bonding boundary remains visible, but the reinforcements now distribute uniformly, resembling the microstructure of the parent composites. Simultaneously, defects such as gas pores and microcracks, commonly associated with the stir-cast method, disappear. This improvement in microstructural homogeneity contributes to a significant increase in the strength of the joint, reaching approximately 55–65% of the strength of the AMCs.

The fracture characteristics of joints welded at various temperatures are depicted in Fig. 3.2. At a relatively low welding temperature of 470 °C, a brittle fracture is observed, characterized by numerous uncovered SiC particles and prominent micropores and cracks. These defects contribute to the low strength of the joint. As the welding temperature increases to 490 °C, the fracture characteristics undergo a noticeable change. The appearance of a liquid phase reduces the number of micropores and cracks visibly, leading to a distinct enhancement in the strength of the welded joints. At a welding temperature of 510 °C, the fracture transitions into a ductile fracture mode, characterized by the absence of uncovered SiC particles and an increase in joint strength. This ductile fracture mode indicates improved plastic deformation and energy absorption capacity, contributing to the enhanced mechanical properties of the joint.

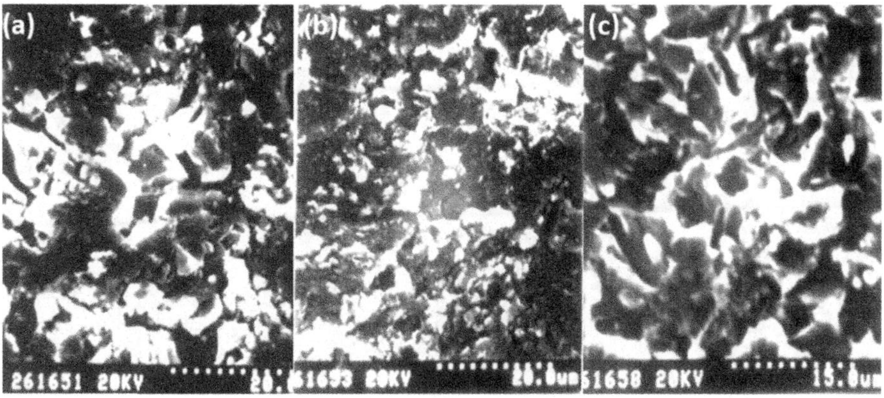

Fig. 3.2 Fracture of the joints diffusion welded at (**a**) 470 °C, (**b**) 490 °C, and (**c**) 510 °C

3.2.2 *Diffusion Welding with Interlayer*

Introducing an interlayer into the weld presents an alternative option for welding AMCs. When selecting an interlayer material, it's essential to consider its compatibility with aluminum and its ability to facilitate diffusion welding. Copper (Cu) and nickel (Ni) are excellent candidates for interlayers in AMC diffusion welding.

3.2.2.1 Cu Interlayer

The eutectic temperature between Cu and Al is widely recognized to be 548 °C [42]. Experimental results indicate that the strength of welded joints can reach up to 132 MPa (approximately 77.6% of the parent composites) under specific welding conditions: a temperature of 550 °C, pressure of 5 MPa, and duration of 20 min, with a copper interlayer thickness of 20 μm. According to the phase diagram, the eutectic reaction does not occur below 540 °C, and the atom diffusion capability is insufficient for successful bonding between the interlayer and the composite. However, at 550 °C, the eutectic reaction occurs, improving the interface state between the reinforcements and changing the joining mode from direct reinforcement–reinforcement contact to reinforcement–interlayer–reinforcement. Additionally, the liquid phase fills micro-holes at the interface, accelerating atom diffusion and facilitating successful joint formation. The correlated fractography, as shown in Fig. 3.3a, reveals that with the increased presence of intermetallic compounds among crystal boundaries, joint strength decreases. Moreover, the parent composites begin to melt, leading to significant joint deformation and poor properties (Fig. 3.3b). At 570 °C, the AMCs melt, increasing mobility and causing particle reinforcements to aggregate in the joint. Brittle eutectic compounds mainly exist among crystal boundaries, distributed in continuous flakes as shown in Fig. 3.4. These factors contribute to lower joint strength, measuring only 38 MPa, equivalent to 22.4% of the strength of the parent composites.

The thickness of the Cu interlayer is a critical parameter in diffusion welding of AMCs. If the thickness of the interlayer is too thin, there is not enough eutectic liquid phase to wet the composite and fill the micro-holes in the welded joint. However, if the thickness of the interlayer is too thick, the Cu element may not diffuse entirely, resulting in the formation of a "no-reinforcement zone" defect in the welded joint. Furthermore, it should be noted that the thickness of the interlayer must match the size of the reinforcement.

Figure 3.5 shows the electron probe microanalyzer (EPMA) micrograph of the welded joint with a 5 μm thickness of Cu interlayer. Figure 3.5a shows the backscattering micrograph of the center of the welded joint, highlighting the Cu element. In Fig. 3.5b, the distribution map of the Cu element is depicted, with the highlighted areas indicating the presence of Cu.

The observation regarding the lower strength of the welded joint due to the thinness of the Cu interlayer is significant. When the interlayer thickness is insufficient, the formation of brittle intermetallic compounds such as $CuAl_2$ may be limited.

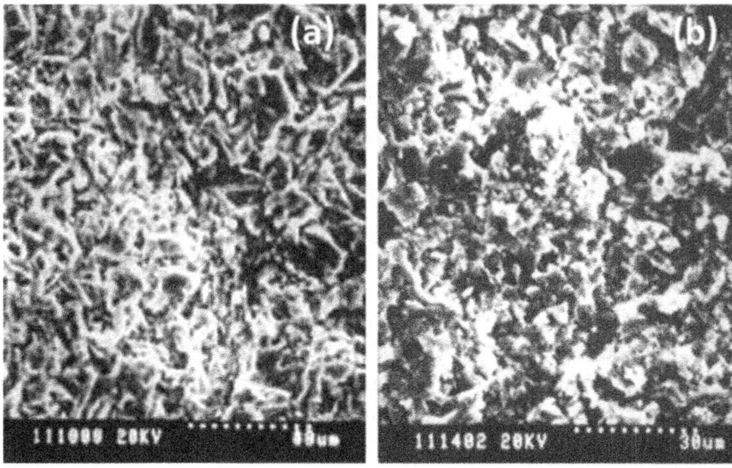

Fig. 3.3 Fracture surface of the welded joints with Cu interlayer at welding temperatures of (**a**) 550 °C and (**b**) 565 °C

Fig. 3.4 EPMA image of the welded joint with Cu interlayer at a welding temperature of 570 °C

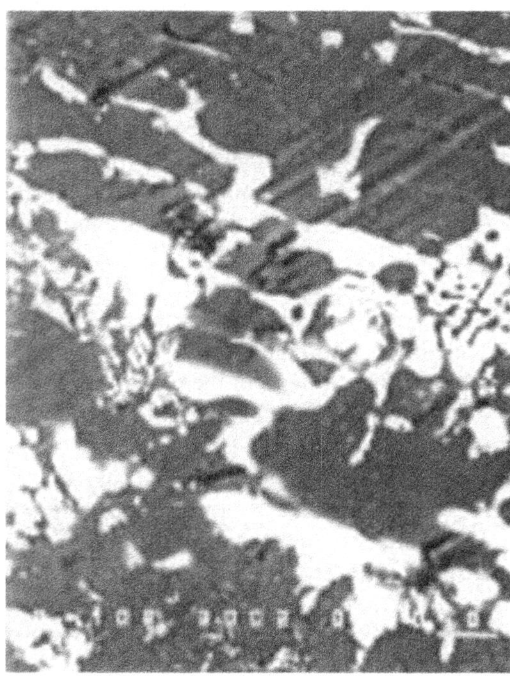

These compounds are crucial for strengthening the joint and enhancing its mechanical properties. The sparse distribution of $CuAl_2$ among the crystal boundaries indicates inadequate bonding and reinforcement within the joint structure. As a result, the joint may exhibit reduced strength and compromised structural integrity.

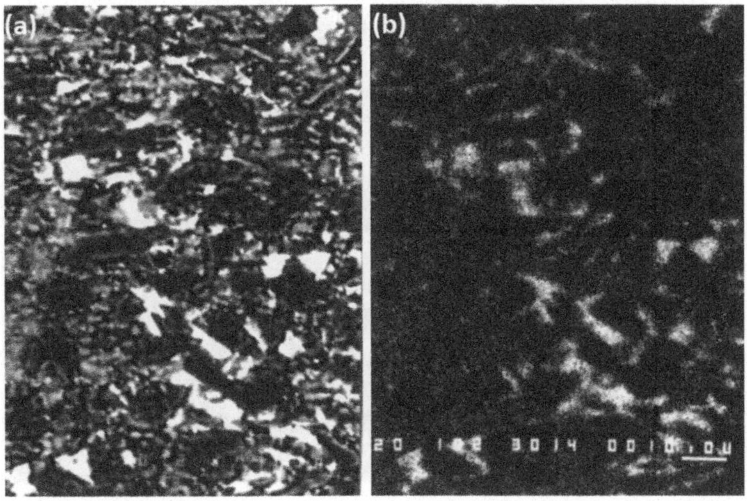

Fig. 3.5 (**a**) Back-scattering micrograph of the welded joint and (**b**) Cu element map distribution for a 5 μm thickness of Cu interlayer

With the increase in the thickness of the Cu interlayer, the strength of the welded joint significantly improves due to the increment in the eutectic liquid phase. Figure 3.6 shows the EPMA micrograph of the welded joint with a 20 μm interlayer thickness. It shows that the intermetallic compound $CuAl_2$ is uniformly distributed among the crystal boundaries, leading to an improved contact state between the reinforcement and matrix through beneficial wetting by the intermetallic compound $CuAl_2$. Simultaneously, the bonding mode shifts from direct reinforcement–reinforcement contact to reinforcement–Cu/Al alloy–reinforcement due to the infiltrating effect. Additionally, the presence of microcracks decreases as the Cu/Al alloy fills the micro-holes in the welded joint. These contributions to the welded joints result in a strength of approximately 132 MPa, which is 77.6% of the strength of the parent composites. If the thickness of the Cu interlayer is too thick (approximately 40 μm), a "no-reinforcement zone" appears in the welded joint, as shown in Fig. 3.7. This "no-reinforcement zone" significantly diminishes the properties of the welded joint.

3.2.2.2 Ni Interlayer

Solidus diffusion welding of stir-cast SiC_p/A356 with a Ni interlayer exhibits similarities to welding with a Cu interlayer. Figure 3.8 shows SEM micrographs of the welded joints at different welding durations, offering insights into the welding process and joint morphology. These micrographs help in understanding the evolution of the welded joints over time and assessing the quality of the joining interface. The comparison between the microstructures of joints welded with Ni interlayers and

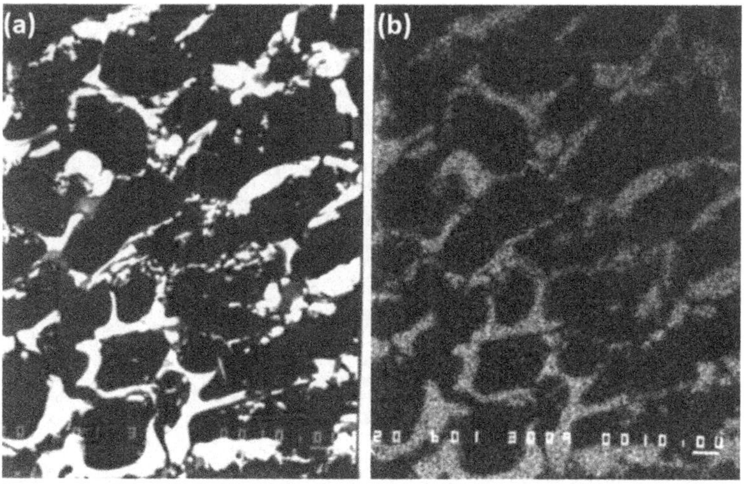

Fig. 3.6 (a) Back-scattering micrograph of the welded joint and (b) Cu element map distribution for a 20 μm thickness of Cu interlayer

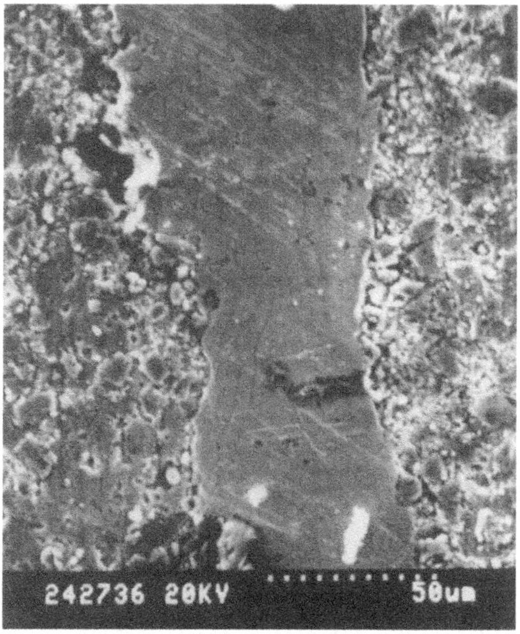

Fig. 3.7 SEM micrograph of the welded joint with a 40 μm thickness of Cu interlayer

those with Cu interlayers can provide valuable information regarding the effectiveness of Ni as an interlayer material and its impact on joint formation and mechanical properties. Analyzing the SEM micrographs allows for the evaluation of factors such as interlayer diffusion, intermetallic compound formation, interface morphology, and the presence of defects or discontinuities within the welded joint.

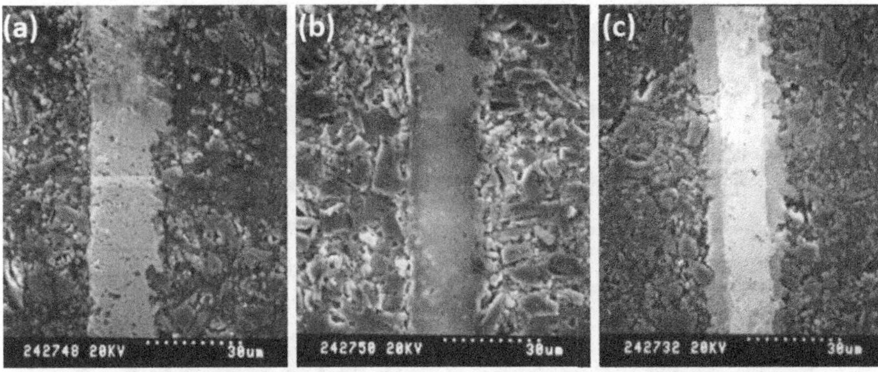

Fig. 3.8 Joints after welding duration of (**a**) 15 min, (**b**) 30 min, and (**c**) 60 min (T = 560 °C, P = 5 MPa, δ = 14 μm)

It indicates that when the welding duration is 15 min, Ni atoms do not apparently diffuse. When the welding duration is 30 min, Ni atoms diffuse over a short distance. Micro-holes exist at the interface between the matrix and interlayer, resulting in the low strength of the welded joint. When the welding duration is 60 min, Ni atoms obviously diffuse, forming a diffusion zone, with the strength of the welded joint reaching up to 121 MPa.

The fractography images are shown in Fig. 3.9, which align well with the results presented in Fig. 3.8. When the welding duration is too short, such as 15 min, the interfaces between the reinforcement and matrix, as well as between reinforcements themselves, have not been entirely joined. This is evidenced by the presence of many SiC particles distributed on the fractured surface, as depicted in Fig. 3.9a. As the welding duration increases to 30 min, the presence of SiC particles on the fractured surface gradually decreases, as shown in Fig. 3.9b. With longer welding durations, such as 60 min, Ni atoms diffuse more sufficiently, leading to a prominent improvement in the contacting areas between the reinforcement–matrix and reinforcement–reinforcement. Consequently, the fracture features change from brittle fracture to plastic fracture with tearing edges, as observed in Fig. 3.9c. However, if the welding duration is extensively prolonged to 90 min, some parts of the parent composites may melt, leading to the successive aggregation of reinforcements and resulting in decreased properties of the welded joints, as shown in Fig. 3.9d.

3.3 Laser Welding

3.3.1 Laser Welding Without Filler

As another conventional joining method and an attractive mature technique, laser welding is widely used for processing composite materials. In this study, stir-cast SiC_p/A356 composites were directly welded using a laser system: a GSI Lumonics

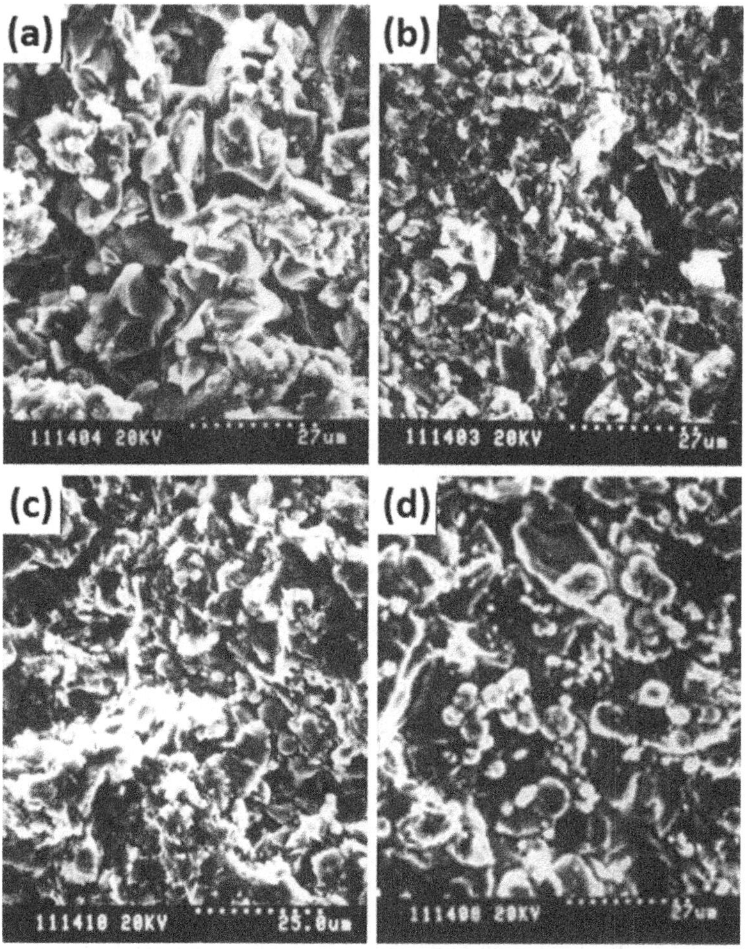

Fig. 3.9 Fractography images for welding duration of (**a**) 15 min, (**b**) 30 min, (**c**) 60 min, and (**d**) 90 min ($T = 560$ °C, $P = 5$ MPa, $\delta = 14$ μm)

Model JK702H Nd:YAG TEM00 mode. The correlated wavelength was 1.06 μm, with a defocused distance of 10 mm resulting in a focus spot diameter of 1.26 mm. The laser energy was set to 2 J, with a frequency of 25 Hz and a pulse duration of 4 ms. Additionally, the welding speed was 300 mm/min.

The microstructure in Fig. 3.10 shows the weld directly produced by laser welding. It reveals the presence of numerous harmful acicular compounds, specifically Al_4C_3, synthesized during the laser welding process. These compounds intertwine extensively, contributing to the poor tensile strength of the weld. The tensile strength of the weld is only about 37.9% (91 MPa) of the strength exhibited by the relevant parent composites. In Fig. 3.11, the correlated fractography further reveals the presence of synthesized compounds, notably Al_4C_3, prominently distributed on the

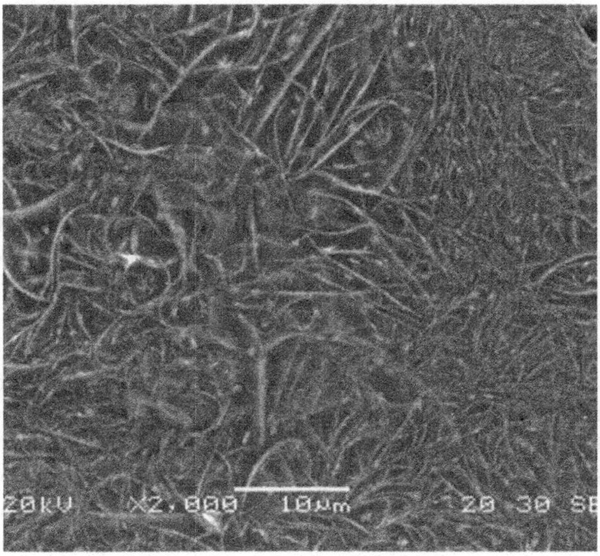

Fig. 3.10 Microstructure of the laser-welded joint without filler

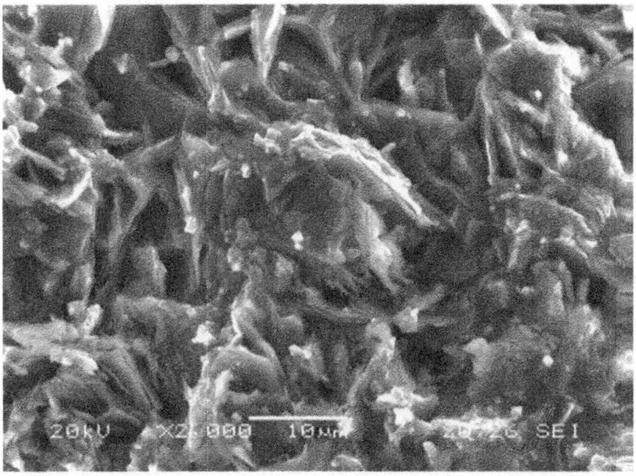

Fig. 3.11 Fractography image of the laser-welded joint without filler

fractured surface. This observation aligns well with the microstructure depicted in Fig. 3.10, confirming the detrimental impact of these compounds on the mechanical properties of the weld. The presence of extensive acicular compounds highlights the challenges associated with laser welding in aluminum matrix composites. These compounds contribute to reduced tensile strength and may compromise the integrity and performance of welded joints.

3.3.2 Laser Welding with Ti Filler

Figures 3.10 and 3.11 reveal that the chemical reaction (4Al + 3SiC → Al_4C_3 + 3Si) occurring during laser welding results in the formation of harmful brittle compounds, particularly Al_4C_3. These compounds significantly degrade the mechanical properties of the welded joints, rendering them unsuitable for practical applications. To address this issue and enhance the properties of the welded joints, the incorporation of titanium alloy, chosen as an ideal filler material, is proposed. By introducing titanium as a filler material, the aim is to mitigate the detrimental chemical reactions during welding and subsequently improve the mechanical properties of the welded joints. Titanium's ability to inhibit the formation of harmful compounds is anticipated to prevent the degradation of joint strength and durability. This approach seeks to enhance the integrity and performance of the welded joints, making them more suitable for practical applications across various industries.

The microstructure in Fig. 3.12 shows the joint welded by laser with a 0.3 mm thick Ti filler. Notably, the brittle compound Al_4C_3, which was observed in previous welds without Ti filler, is absent in this weld. Instead, newly synthesized phases of TiC are uniformly distributed throughout the weld. Furthermore, the composite material fully melts during the laser welding process. As a result, the synthesized TiC particles, acting as reinforcements, not only distribute uniformly in the weld but also exhibit a more compact distribution. This optimized distribution of TiC particles contributes to the enhanced mechanical properties of the welded joint. The absence of Al_4C_3 and the presence of TiC particles indicate that the addition of a titanium filler effectively suppresses the formation of detrimental compounds and promotes the synthesis of particles that contribute to the mechanical integrity of the weld. This improvement suggests that laser welding with a Ti filler holds promise for enhancing the properties and performance of welded joints in AMCs.

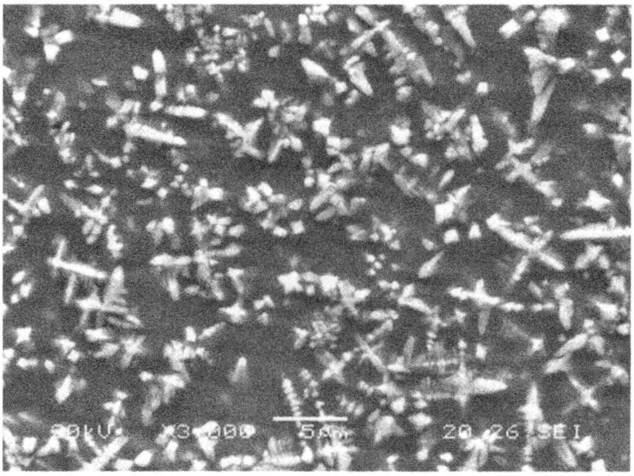

Fig. 3.12 Microstructure of the weld with a 0.3 mm thick Ti filler

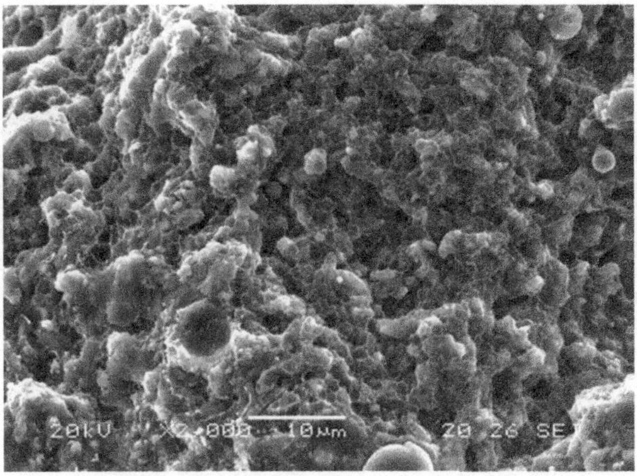

Fig. 3.13 Fractography image of the weld with a 0.3 mm thick Ti filler

Figure 3.13 presents the fractography corresponding to the microstructure depicted in Fig. 3.12. The fractured surface exhibits a large number of dimples, along with spherical TiC particles distributed throughout. This observation indicates a significant shift in the fracture behavior of the weld, transitioning from brittle fracture to ductile fracture, particularly when compared to direct laser welding processes. The presence of dimples and spherical TiC particles suggests improved ductility and enhanced mechanical properties of the welded joint. This improvement can be attributed to the beneficial in-situ reaction between SiC particles and titanium during laser irradiation (Ti + SiC → TiC + Si). This reaction facilitates the replacement of the SiC/Al interface with TiC/Al, leading to improved interfacial contact and enhanced flowability of the welding pool. Additionally, the in-situ synthesized TiC particles exhibit enhanced distribution within the weld, facilitated by the stirring effect during the welding process. The effective suppression of harmful compounds such as Al_4C_3 at the interface between aluminum and SiC further underscores the feasibility and effectiveness of using titanium as a filler material for laser welding of AMCs.

Figure 3.14 shows the influence of titanium on the microstructure evolution with varying thicknesses, while the corresponding fractograph is depicted in Fig. 3.15. Overall, the presence of titanium contributes significantly to improvements in the properties of the welded joints, particularly with increasing titanium thickness. In Fig. 3.14a, where the titanium thickness is 0.15 mm, sporadic scattering of the in-situ synthesized particles (TiC) is observed. However, a large quantity of deleterious acicular Al_4C_3 still forms and is substantially distributed within the weld. This indicates that the Al_4C_3 compound is not effectively restrained during the laser welding process. Conversely, when the titanium thickness exceeds 0.3 mm, the tensile strength of the weld decreases due to the failure of the in-situ reaction. Figure 3.14b illustrates the microstructure of the weld with a titanium thickness of 0.45 mm, showing a poorer microstructure compared to the 0.3 mm thickness. With increased

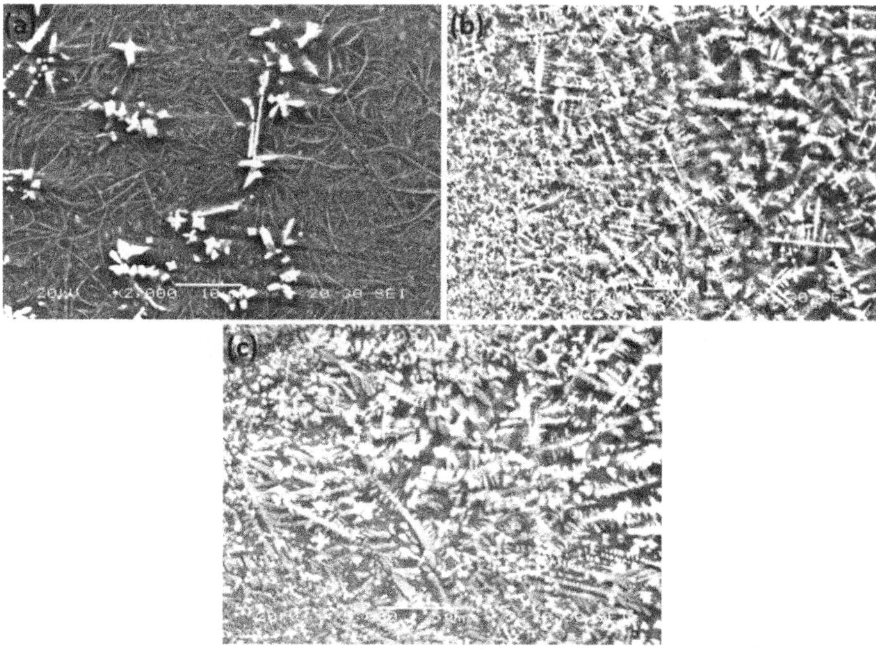

Fig. 3.14 Microstructure of welds with Ti filler thickness of (**a**) 0.15 mm, (**b**) 0.45 mm, and (**c**) 0.60 mm

titanium thickness, a larger portion of the energy irradiated by the laser is absorbed by the titanium, resulting in less energy available for the welding process.

As the thickness of the titanium filler increases, the energy input into the AMCs decreases correspondingly, affecting the occurrence of the beneficial in-situ reaction during welding. The associated microstructure is depicted in Fig. 3.14b, where fine equiaxed crystals and coarse columnar crystals coexist within the weld, leading to a reduction in welding properties. With further increases in titanium thickness, an even greater amount of energy irradiated by the laser is required to melt the titanium. Consequently, the energy input into the AMCs diminishes relatively. This restriction significantly impedes the occurrence of the in-situ reaction within the welding pool. Consequently, a greater number of coarse columnar crystals are detected, accompanied by scattered fine equiaxed crystals, as illustrated in Fig. 3.14c.

The fracture surface reveals the presence of some bare SiC particles, along with the appearance of harmful Al_4C_3, as shown in Fig. 3.15a. In the case of welded joints with a titanium thickness of 0.45 mm, the fracture surfaces have clear indications of fine equiaxed crystals and coarse columnar crystals. Additionally, spherical TiC particles are observed to distribute across the fracture surface, as seen in Fig. 3.15b. These characteristics in the microstructure evolution consequently contribute to the diminished properties of the welded joints. In the case of welded joints with a

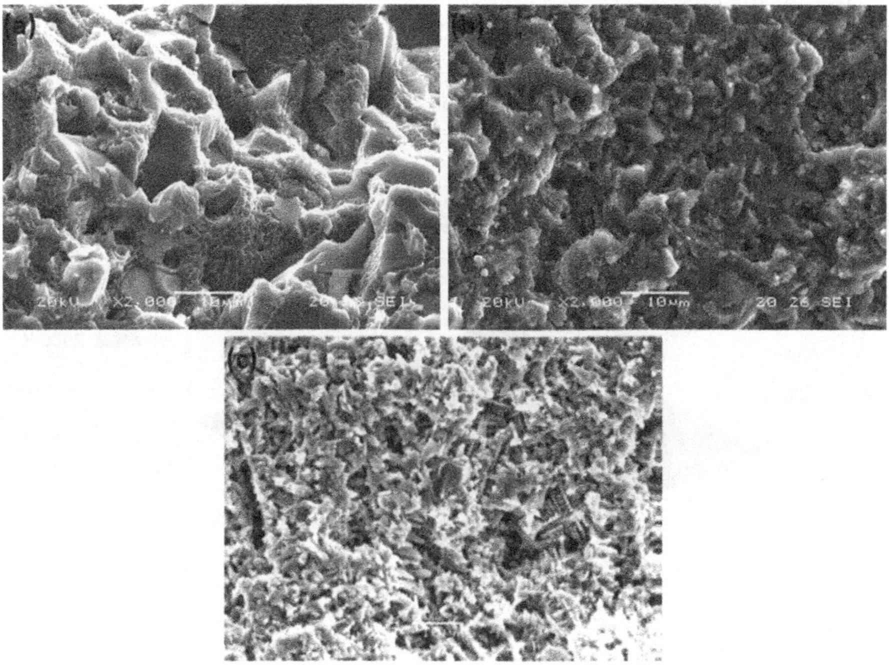

Fig. 3.15 Fractography images with Ti filler thickness of (**a**) 0.15 mm, (**b**) 0.45 mm, and (**c**) 0.60 mm

titanium thickness of 0.60 mm, the fracture surface, as shown in Fig. 3.15c, primarily comprises coarse columnar crystals, with only a few scattered fine equiaxed crystals. This observation aligns well with the results depicted in Fig. 3.14c, indicating a predominance of coarse columnar crystals and limited fine equiaxed crystals in the microstructure evolution.

The calculated laser welding temperatures are shown in Figs. 3.16 and 3.17. It is evident that the temperature field of laser welding AMCs without titanium remains consistent with the conventional approach. In contrast, with the introduction of titanium, the maximum temperature increases due to the energy required to melt the titanium. However, the temperature distribution becomes more uniform due to the endothermic reaction and in-situ synthesis associated with titanium melting. This results in a smoother temperature distribution compared to direct laser welding, along with a reduction in the width of the heat-affected zone (HAZ).

The phase diagram of the Ti-Al binary system, coupled with the anticipated welding temperatures, suggests that variations in titanium content in the welding pool can lead to the formation of brittle intermetallic compounds such as TiAl and Ti_3Al. This occurs due to reactions involving Ti, Al, and SiC particles, ultimately resulting in the formation of TiC and various intermetallic compounds. Consequently,

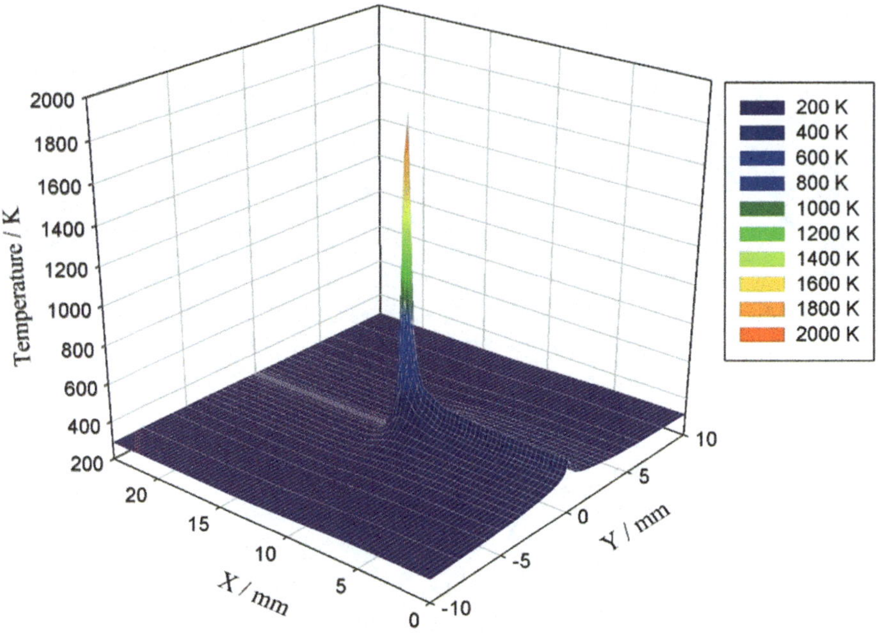

Fig. 3.16 Temperature distribution without Ti filler

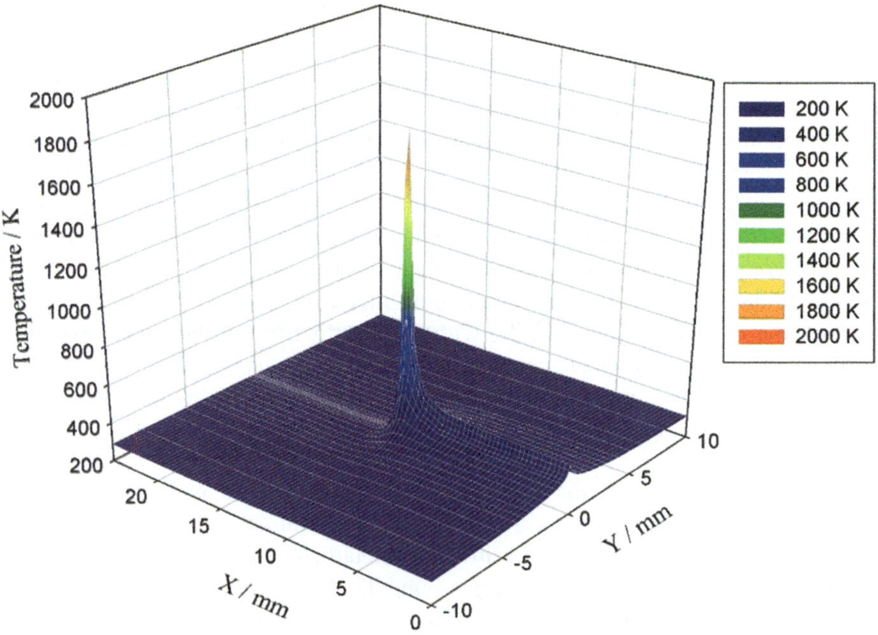

Fig. 3.17 Temperature distribution with Ti filler

the presence of these compounds can adversely affect the mechanical properties of the welded joints. When the thickness of the titanium interlayer is excessive, the beneficial in-situ reaction leading to TiC nucleation may be suppressed. Instead, large blocks of columnar crystals may form, which can ultimately compromise the integrity of the weld joint.

In summary, as the thickness of the titanium interlayer increases, the amount of energy irradiated into the AMCs decreases, while the energy required to melt the titanium increases. This imbalance can lead to insufficient in-situ reactions and a weakened stirring effect in the welding pool, resulting in lower tensile strength of the welded joints.

3.4 Transient Pulse-Impact Liquid-Phase Welding (TPILPW)

Transient pulse-impact liquid-phase welding (TPILPW) for joining AMCs represents a promising approach to address the limitations encountered with traditional welding methods. In contrast to diffusion welding, which can be time-consuming and inefficient, and laser welding, which may result in the formation of harmful intermetallic compounds, TPILPW offers a potentially more effective solution for joining AMCs.

This innovative welding technique utilizes transient pulses and impacts to create a liquid phase at the interface, facilitating the bonding process. Due to the transient nature of the pulses, the welding process can be completed more efficiently compared to traditional methods [43–46]. Additionally, the formation of a liquid phase helps promote strong bonds between the materials without the formation of detrimental intermetallic compounds. Through the application of TPILPW, challenges associated with conventional welding techniques, such as weak tensile strength and the formation of brittle compounds, can be mitigated. This approach holds promise for enhancing the weld quality and efficiency in joining AMCs, thereby making it more suitable for practical applications in various industries.

The basic operation of TPILPW is shown in Fig. 3.18. It is important to highlight that the operating temperature during TPILPW falls between the solid and liquid phases of AMCs. During the process, pulse impacts are applied to the samples within a short duration ranging from 10^{-4} to 10^{-2} s, facilitating the anticipated deformation and enabling the welding operation. This short duration of pulse impacts ensures that the welding process occurs swiftly and efficiently. By carefully controlling the duration and intensity of the pulse impacts, TPILPW creates localized heating at the interface of the materials, inducing a liquid phase that promotes bonding between the AMCs. This liquid phase facilitates the formation of strong and durable welds without the formation of harmful intermetallic compounds that could compromise the integrity of the joint.

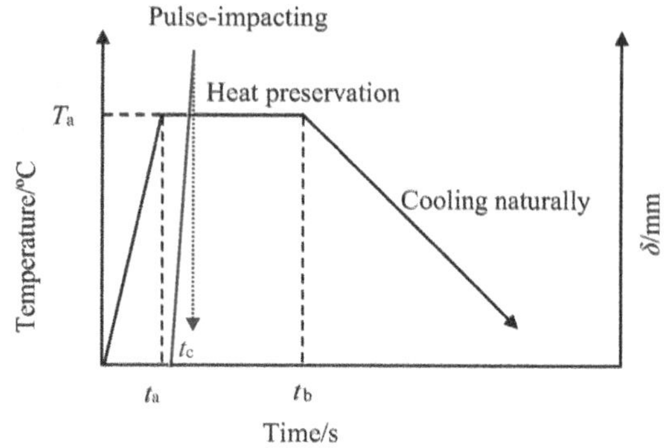

Fig. 3.18 Schematic diagram of transient pulse-impact liquid-phase welding

3.4.1 Pulse-Impact Mode

In the TPILPW process, there are three distinct modes for the application of pulse impacts, each corresponding to different conditions and durations. In Mode I, the impact is applied immediately at time t_{c1} when the temperature T_a is held at 570 °C, with a holding time of less than 5 s. Mode II involves the impact being applied at time t_{c2} when T_a is held at 570 °C, with the holding time ranging from 5 to 10 s. Mode III entails the impact being applied at time t_{c3} when T_a is held at 570 °C, with the holding time exceeding 10 s. The schematic diagram illustrating these modes is presented in Fig. 3.19.

Additionally, images of the fractured surfaces corresponding to each mode are depicted in Figs. 3.20, 3.21, and 3.22. In these images, the welding temperature T is maintained at 570 °C, the impact speed V_I is set to 560 mm/s, the welding time t is 30 s, and the pressure P is maintained at 5 MPa. These parameters collectively define the conditions under which the TPILPW process is carried out, influencing the quality and characteristics of the welded joints.

The analysis indicates that the insufficient amount of liquid phase formed in Mode I leads to sporadic welding of the joint, with SiC particles uncovered by aluminum aggregating and scattering on the fractured surface, as shown in Fig. 3.20. On the other hand, in Fig. 3.21, SiC particles are not detected on the fractured surface, indicating that the amount of liquid phase is sufficient to improve the interface condition between aluminum and SiC in the weld in Mode II, resulting in a prominent fractography. In Mode III, due to the extended holding time, the heated zone enlarges, and the rigidity of AMCs distinctly reduces. Under the effect of pulse-impact, AMCs slide against each other, resulting in the failure surface as shown in Fig. 3.22. It should be noted that the impact speed must be controlled at 560 mm/s. If the speed exceeds 700 mm/s, the formed liquid matrix will be squeezed out of the weld, leading to an insufficient liquid phase to cover SiC particles. Consequently, particles seriously congregate, resulting in a noticeable decline in the properties of the welded joints, as depicted in Fig. 3.23.

Figure 3.24 shows the fracture characteristics of the welded joints at various temperatures, where V_I = 560 mm/s, t_I = 10^{-2} ~ 10^{-4} s, and δ (overall horizontal

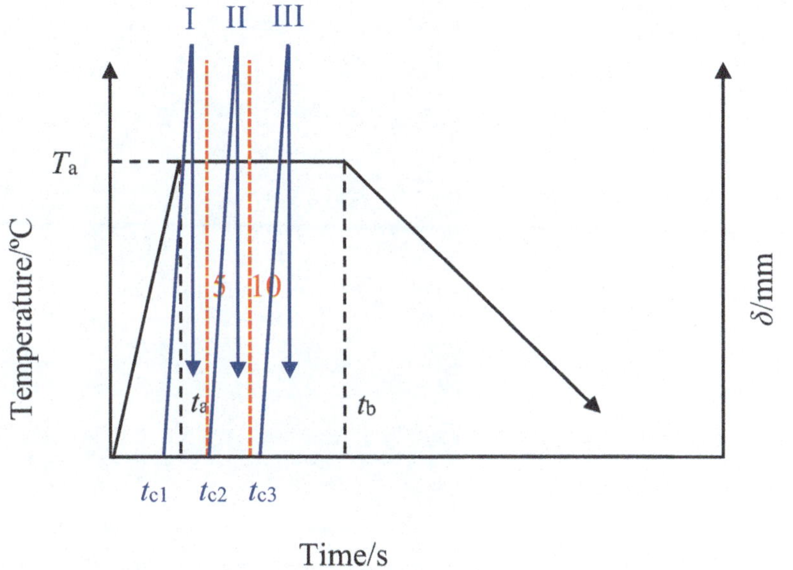

Fig. 3.19 Schematic diagram of different pulse-impact modes

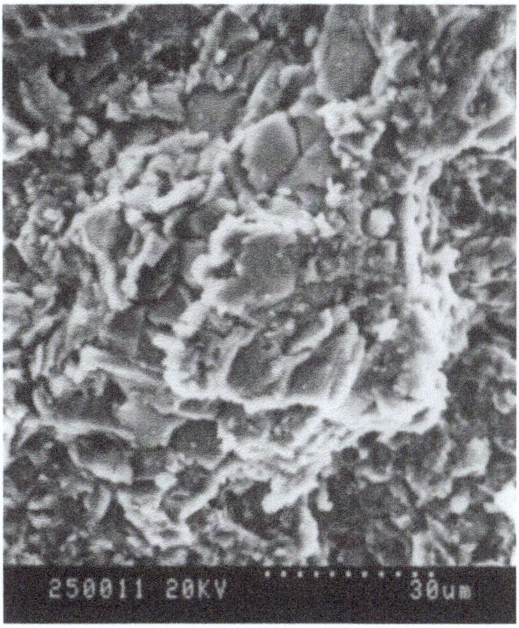

Fig. 3.20 Fractography image of the welded joint by Mode I

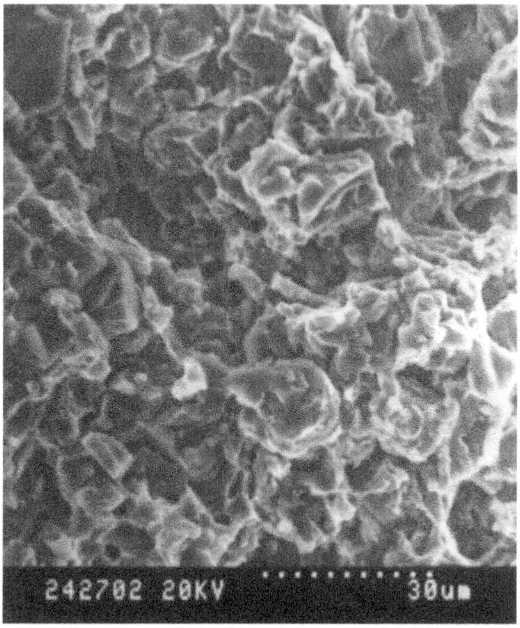

Fig. 3.21 Fractography image of the welded joint by Mode II

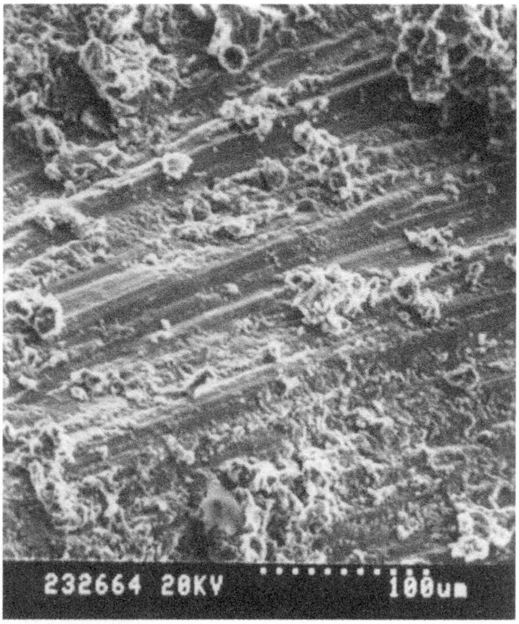

Fig. 3.22 Fractography image of the welded joint by Mode III

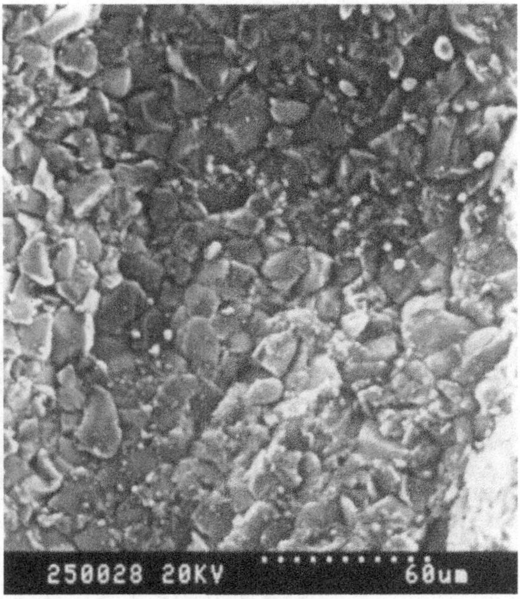

Fig. 3.23 Fractography image of the welded joint at impact speed higher than 700 mm/s

deformation) = 1 mm, according to practical requirements. The parameters set for this analysis include t = 30 s and P = 5 MPa.

At 563 °C, SiC_p/A356-based AMCs sporadically weld with uncovered SiC particles densely distributed in some areas. Moreover, the initial topography of AMCs is still obviously distinguished on the fractured surface, as shown in Fig. 3.24a. At 565 °C, as shown in Fig. 3.24b, owing to the more sufficient formation of liquid phases, the properties of the welded joint are improved. The joint is excellently welded in some locations, and issues related to SiC particle aggregation are ameliorated to some extent. However, there are still some uncovered SiC particles scattered on the fractured surface, indicating unsuccessful AMCs welds. At 570 °C, the characteristic of the fractography is dimple fracture, with some SiC particles distributed in the dimples as shown in Fig. 3.24c. This expresses that the SiC particles as reinforcement are effectively wetted by the aluminum matrix, and no harmful or brittle phases form in the weld, resulting in favorable welded joints. AMCs specimens relatively slide due to the formation of a more liquid phase aluminum matrix and the reduction in AMCs rigidity, resulting in the failure of the weld.

3.4.2 TPILPW Mechanism

Due to the low conductivity and high melting point properties of SiC particles, the temperature of the contact welding area is higher than in other parts. During welding, heat energy severely congregates at the welding area, leading to the formation

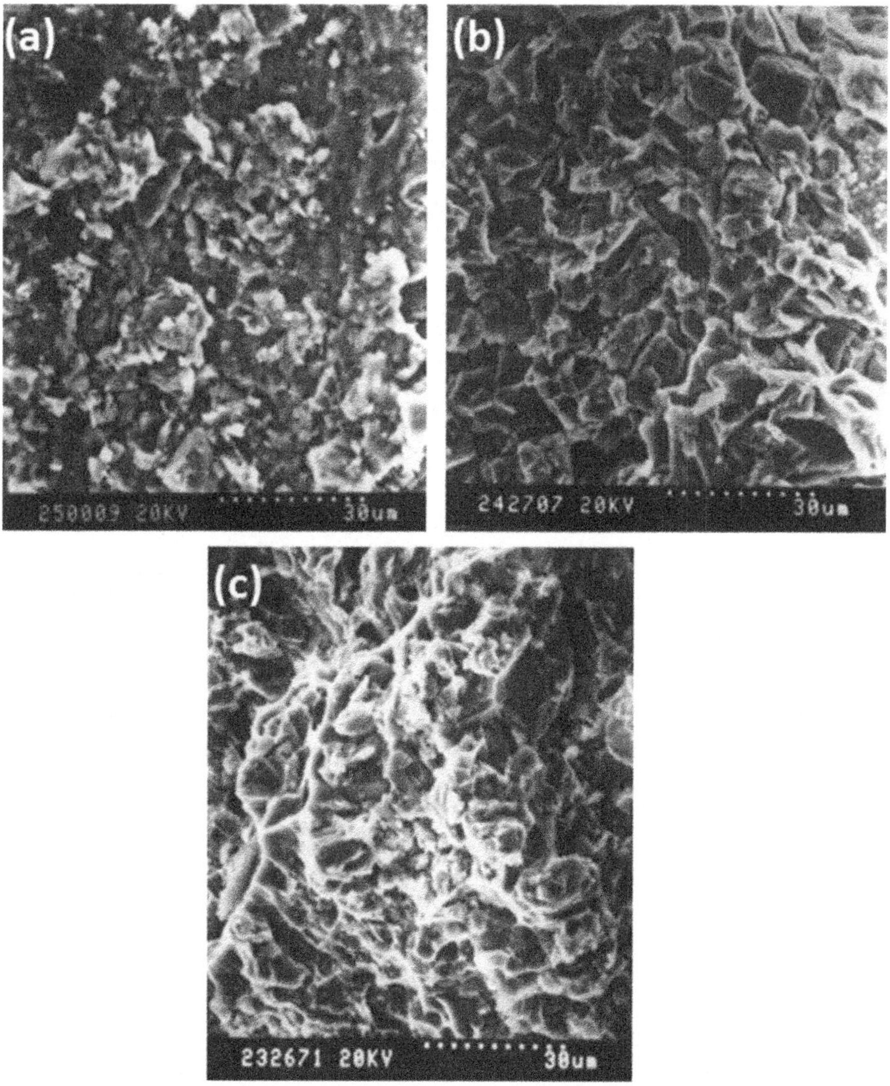

Fig. 3.24 Fractography images of the welded joints at temperatures of (**a**) 563 °C, (**b**) 565 °C, and (**c**) 570 °C

of a sufficient liquid phase aluminum matrix (Fig. 3.25a). Under the effect of transient pulse-impact, the joint is achieved through intensive diffusion, aided by plastic deformation. Subsequently, the additional energy released by transient pulse-impact also promotes the formation of nano-grains in the aluminum matrix, favorably enhancing the strength of the welded joints. Meanwhile, under the squeezing effect of transient pulse-impact, tiny SiC particles and a small amount of liquid phase are pushed out from the welding zone (Fig. 3.25b), facilitating the cleaning off of the

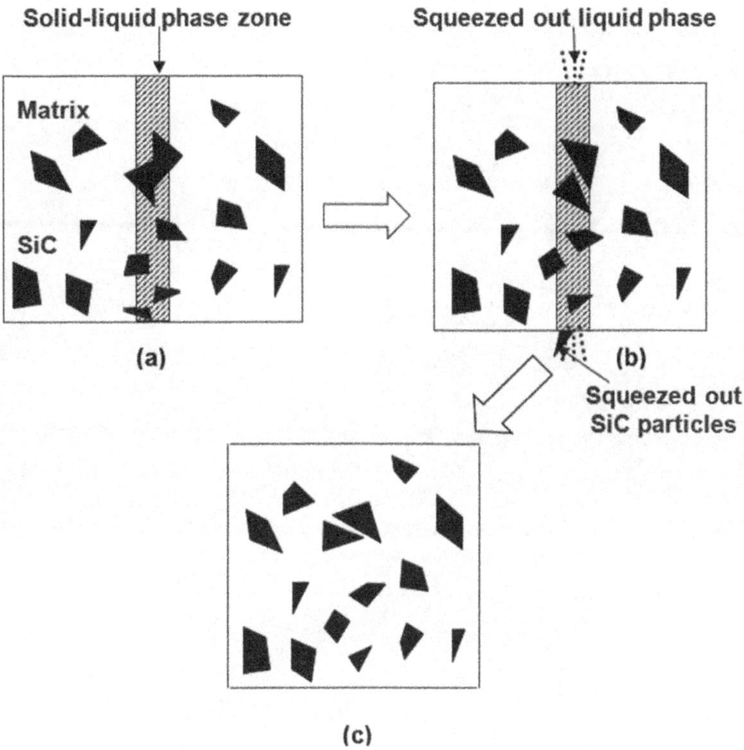

Fig. 3.25 Schematic representation of the mechanism of transient pulse-impact liquid-phase welding

remnant oxide film. Additionally, the flow of the matrix forces SiC particles to flow synchronously, resulting in a more uniform distribution of SiC particles. Furthermore, it should be noted that the transient pulse-impact beneficially eliminates interior apertures dispersed in the initial AMCs, modifying and improving the density of AMCs to some degree (Fig. 3.25c).

3.4.3 Fracture Characteristics of SiC_p/AA6061 and Al_2O_{3p}/AA6061 Welded by TPILPW at Various Temperatures

To determine the feasibility of TPILPW, powder metallurgy-based SiC_p/AA6061 and Al_2O_{3p}/AA6061 AMCs are employed for validation investigation. These composites are reinforced with 15% volume fraction SiC and Al_2O_3 particles of 5 μm mean size.

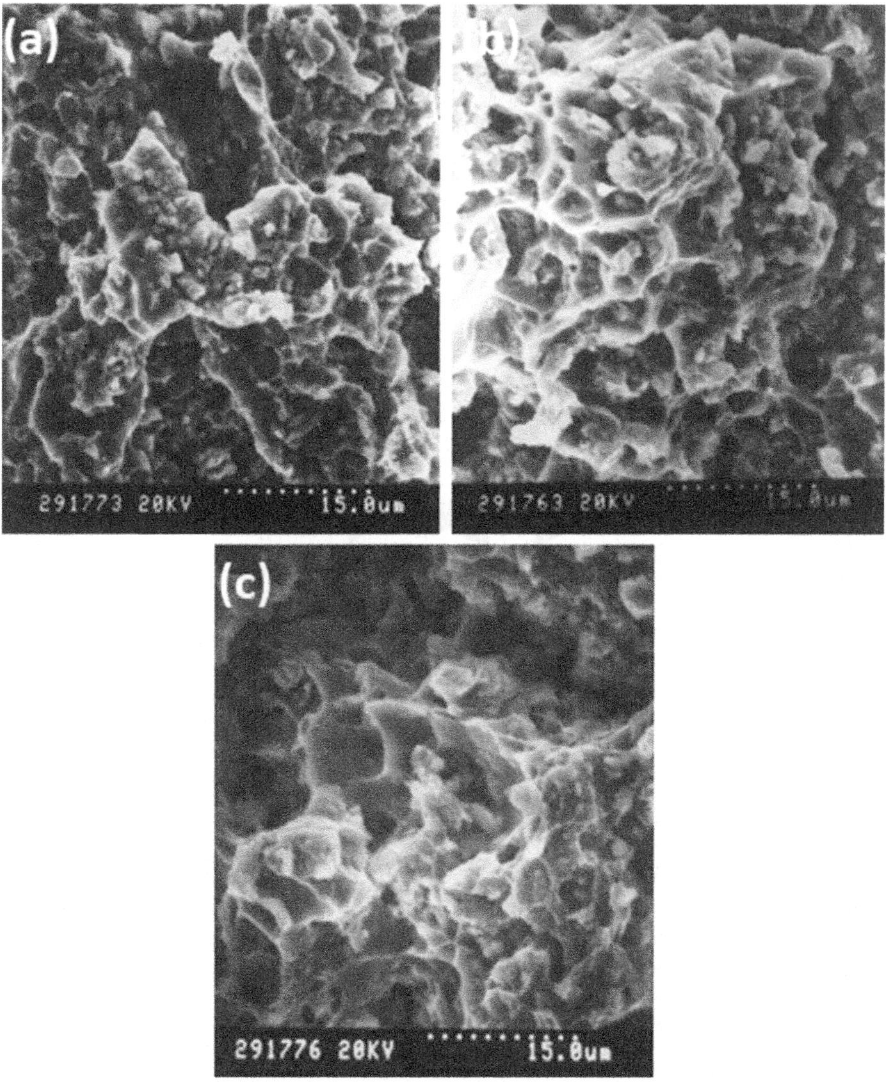

Fig. 3.26 Fractography images of SiC_p/AA6061 at temperatures of (**a**) 620 °C, (**b**) 623 °C, and (**c**) 625 °C

Figures 3.26 and 3.27 show the fracture characteristics of SiC_p/AA6061 and Al_2O_{3p}/AA6061 processed by TPILPW at various temperatures. It is observed that the fractured surfaces under the effect of transient pulse-impact are similar to those of SiC_p/A356. They exhibit dimple fractures with some reinforcement particles (SiC, Al_2O_3) embedded in the dimples. This suggests that the welded joints of AMCs can be easily achieved with high strength using the proposed approach. Moreover, owing to the prominent effect of pulse-impact, the weld can be realized in 30 s, which is remarkably shorter than conventional methods.

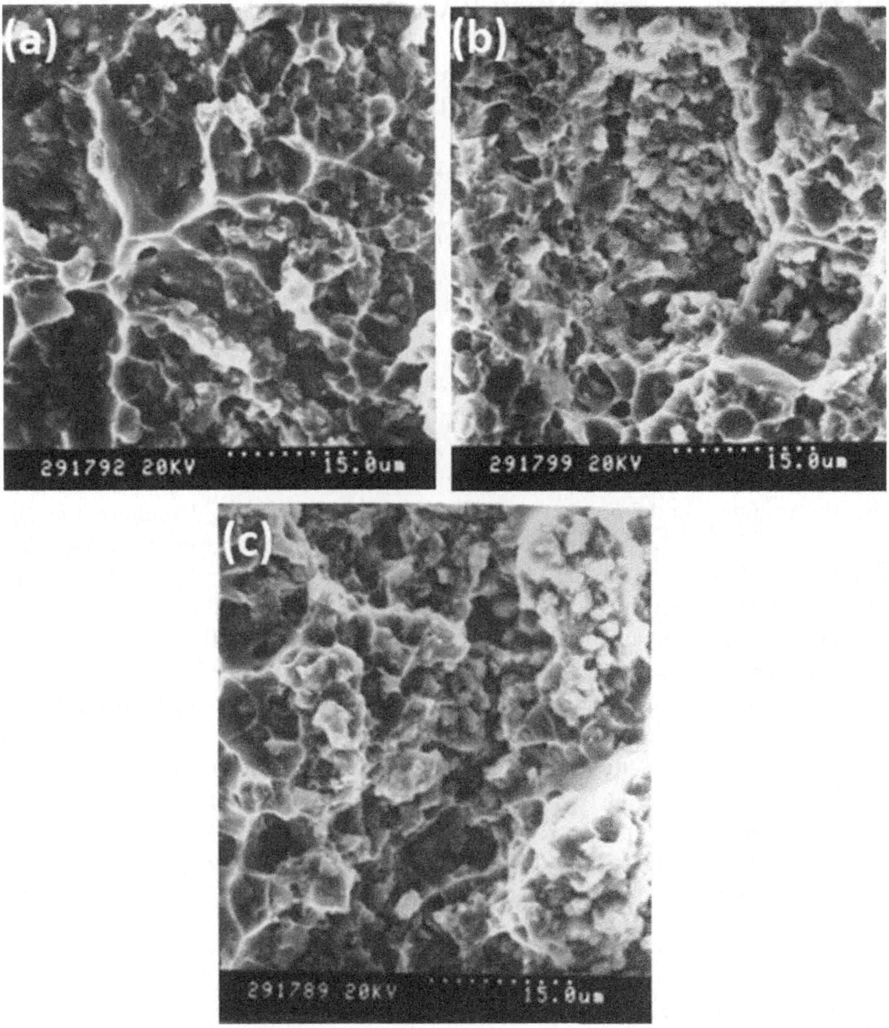

Fig. 3.27 Fractography images of Al_2O_{3p}/AA6061 at temperatures of (**a**) 641 °C, (**b**) 644 °C, and (**c**) 647 °C

While the results of TPILPW are captivating, it is important to acknowledge some drawbacks associated with this technique. These include limited options for welding temperature, challenges in welding complex structures, and the technique not being a near-net-shape technique, among others [47–49]. Furthermore, in line with eco-requirements and the increasing need to reduce environmental footprints and various wastes, there is a growing emphasis on exploring green MMCs. This exploration aims to minimize our environmental impact and address environmental concerns [50–53].

3.4.3.1 Interface in TPILPW $SiC_p/AA6061$

Equation 3.1 represents the reaction where SiC reacts with Al to form Al4C3, which could occur at high temperatures during TPILPW for $SiC_p/AA6061$.

$$3SiC(s) + 4Al(l) \rightarrow Al_4C_3(s) + 3Si(s) \tag{3.1}$$

The correlated free energy [42] is given in Eq. 3.2.

$$\begin{aligned}\Delta G\left(J.mol^{-1}\right) = 113900 - 12.06T\ln T + 8.92\times10^{-3}T^2 \\ +7.53\times10^{-4}T^{-1} + 21.5T + 3RT\ln\alpha_{[Si]}\end{aligned} \tag{3.2}$$

where $\alpha_{[Si]}$ is the silicon activity in the liquid aluminum.

If the change in Gibbs free energy (ΔG) is greater than zero, it indicates that the reaction is non-spontaneous under the given conditions according to the laws of thermodynamics. In the context of TPILPW of $SiC_p/AA6061$, it suggests that the process may not occur spontaneously under those particular welding conditions. However, considering the binary alloy phase diagram of Al-Si, silicon will dissolve and its relevant activity in liquid aluminum evolves at different temperatures. Figure 3.28 shows the evolution of free energy at various TPILPW temperatures, indicating that the reaction given in Eq. 3.1 is able to occur in TPILPW.

It is well-known that even slight reactions at interfaces can significantly enhance the properties of welded joints. This is primarily due to their advantageous effect on the release of internal stress, induced to some extent by hetero-matches between reinforcements (SiC, Al_2O_3) and the matrix. The interface condition, particularly the reinforcements-matrix interface, plays a crucial role in AMCs and is improved by the relevant reactions to modify their mismatches. Figure 3.29 shows the

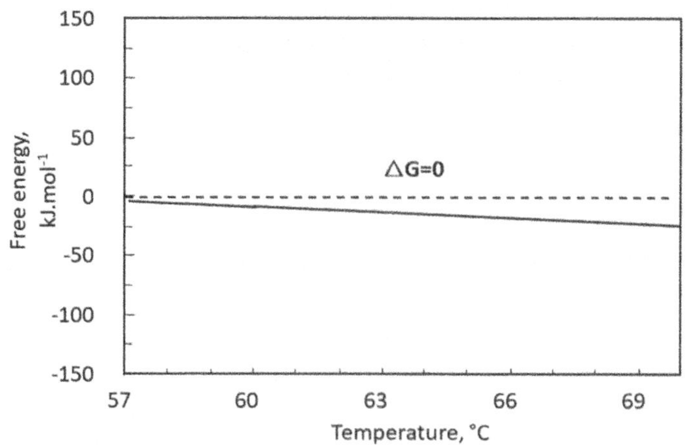

Fig. 3.28 Evolution of free energy at various welding temperatures

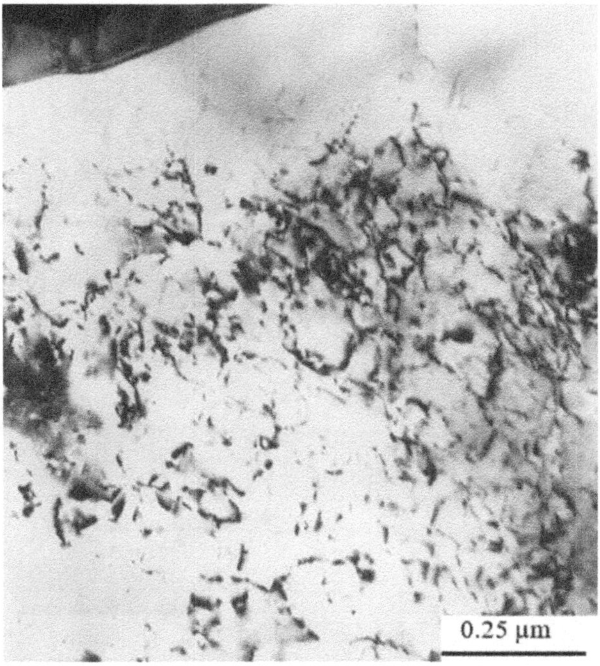

Fig. 3.29 Dislocations at the interface of the welded joint of SiC_p/AA6061

distribution of dislocations at the interface of SiC_p/AA6061 TPILPW. It can be observed that the density of dislocations apparently decreases, demonstrating that the interface reaction occurs and effectively releases the internal stress. Therefore, the welded joints of SiC_p/AA6061 convincingly possess compelling properties better than those of Al_2O_{3p}/AA6061.

3.4.3.2 Interface in TPILPW SiC_p/A356 and Al_2O_{3p}/AA6061

For comparison, Figs. 3.30 and 3.31 show the distribution of dislocations at the interface of the welded joints of SiC_p/A356 and Al_2O_{3p}/AA6061 by TPILPW. In these cases, there does not exist any interface reaction. The clear interface between particles and the matrix is distinctive, along with mutually entwisting dislocations in the matrix. These prominent characteristics ultimately contribute to the relatively superior strength of the welded joint compared to traditional diffusion welding [20, 29, 31, 32].

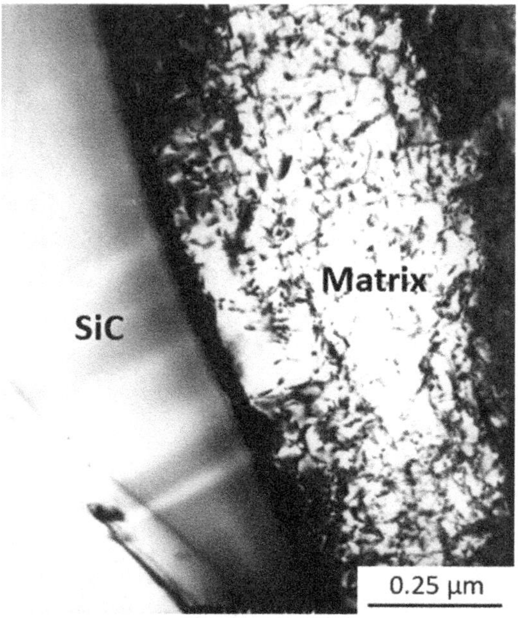

Fig. 3.30 Dislocations at the interface of the welded joint of SiC_p/A356

Fig. 3.31 Dislocations at the interface of the welded joint of Al_2O_{3p}/AA6061

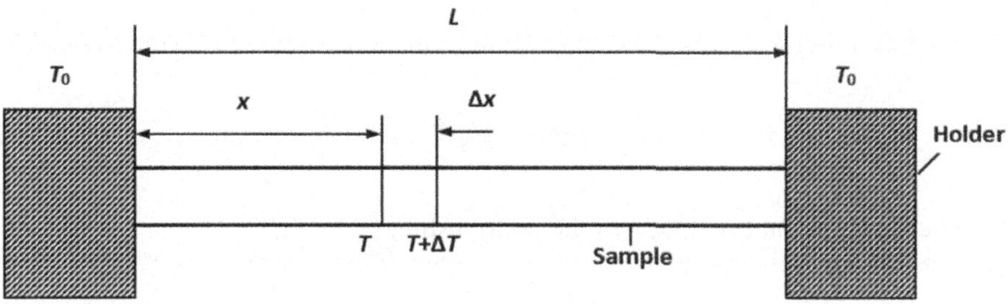

Fig. 3.32 Schematic diagram of temperature distribution in a heated sample

3.5 Calculation of Temperature Distribution

Figure 3.32 shows the schematic of temperature distribution in a heated sample, where L represents the length of the sample between two holders. When the sample is heated, the heat transfer at a distance of x in a time interval of Δt is given by Eq. 3.3:

$$Q = -\alpha A \frac{\partial T}{\partial x} \Delta t \tag{3.3}$$

where T is the temperature, A is the cross-section area, and α is the coefficient of thermal conductivity.

Equation 3.4 gives the heat transfer at the distance of $(x + \Delta x)$:

$$Q + \Delta Q = -\alpha A \left[\frac{\partial T}{\partial x} + \frac{\partial}{\partial x}\left(\frac{\partial T}{\partial x}\right)\Delta x \right] \Delta t \tag{3.4}$$

Neglecting the loss due to convection and irradiation, the energy obtained per unit length at a distance of Δx is calculated by Eq. 3.5:

$$\Delta Q + WA\Delta x\Delta t = \alpha A \frac{\partial^2 T}{\partial x^2} \Delta x \Delta t \tag{3.5}$$

where W is the heat input per unit volume.

Therefore, the rate of temperature variation is given by Eq. 3.6:

$$\frac{\partial T}{\partial x} = \frac{1}{\rho c}\left(W + \alpha \frac{\partial^2 T}{\partial x^2} \right) \tag{3.6}$$

where c is the specific heat capacity, and ρ is the density of the sample.

Once the processing is stable, $\frac{\partial T}{\partial x} = 0$, the equation becomes as given in Eq. 3.7:

$$W = -\alpha \frac{\partial^2 T}{\partial x^2} \tag{3.7}$$

After integration, Eq. 3.8 gives the relationship:

$$\frac{\partial T}{\partial x} = -\frac{W}{\alpha} x + b \tag{3.8}$$

At $x = \frac{1}{2} L$, and $\frac{\partial T}{\partial x} = 0$, the equation becomes as given in Eq. 3.9:

$$b = \frac{WL}{2\alpha} \tag{3.9}$$

After integrating Eq. 3.8, and applying Eq. 3.9, the relationship becomes as given in Eq. 3.10:

$$T = -\frac{W}{2\alpha} x^2 + \frac{WL}{2\alpha} x + b' \tag{3.10}$$

$T = T_0$ at $x = 0$. Hence, $b' = T_0$, and Eq. 3.10 turns into Eq. 3.11:

$$T = -\frac{W}{2\alpha} x^2 + \frac{WL}{2\alpha} x + T_0 \tag{3.11}$$

The relationship can be written as given in Eq. 3.12:

$$T - T_0 = \frac{W}{2\alpha} (L - x) x \tag{3.12}$$

$T = T_{max}$ at $x = \frac{1}{2} L$. Hence, the equation becomes as given in Eq. 3.13:

$$T_{max} - T_0 = \frac{WL^2}{8\alpha} \tag{3.13}$$

Suppose that $T = T_{max} - \Delta T$. Hence, the relationship turns into Eq. 3.14 by combining Eqs. 3.12 and 3.13:

$$T_{max} - \Delta T - T_0 = 4 \frac{(T_{max} - T_0)}{L^2} (L - x) x \tag{3.14}$$

Then, the distance x can be found by Eq. 3.15:

$$x = \frac{L}{2}\left[1 \pm \sqrt{\frac{\Delta T}{T_{max} - T_0}}\right] \tag{3.15}$$

Consequently, the length between $T_{max}$and $T_{max} - \Delta T$ is calculated by Eq. 3.16:

$$\Delta x = L\sqrt{\frac{\Delta T}{T_{max} - T_0}} \tag{3.16}$$

It should be noted that due to the hetero-matches (mismatches) at the welding area, the resistance between two welded samples is higher than that of the relevant calculation. Therefore, Eq. 3.16 should be modified to Eq. 3.17, where η is the coefficient of the mismatched effect of hetero-resistance at the welding area.

$$\Delta x = \eta L\sqrt{\frac{\Delta T}{T_{max} - T_0}} \tag{3.17}$$

3.6 Conclusion

Due to the discontinuous nature of AMCs, several issues such as reinforcement aggregation or absence (creating no-reinforcement zones), the formation of pernicious phases like Al_4C_3, long welding times, and low efficiency persist with conventional welding methods to achieve high-quality welded joints. Although TPILPW has been explored for AMCs, it still presents some deficiencies. While it offers prominent quality of welded joints and high welding efficiency, it has limitations such as relatively narrow optimal welding temperatures, unsuitability for complex structures, lack of near-net-shape processing capability, and the requirement for preprocessing.

Therefore, efforts are still needed to explore new feasible and simple welding techniques for AMCs or MMCs, especially focusing on practical applications. Moreover, in line with global requirements for reducing carbon emissions, eco-friendly MMCs should be explored to decrease our environmental footprint and minimize various wastes in the environment.

References

1. Kashaev N, Ventzke V, Çam G. Prospects of laser beam welding and friction stir welding processes for aluminum airframe structural applications. J Manuf Process. 2018;36:571–600.
2. Garg P, Jamwal A, Kumar D, Sadasivuni KK, Hussain CM, Gupta P. Advance research progresses in aluminium matrix composites: manufacturing & applications. J Mater Res Technol. 2019;8(5):4924–39.
3. Chak V, Chattopadhyay H, Dora TL. A review on fabrication methods, reinforcements and mechanical properties of aluminum matrix composites. J Manuf Process. 2020;56(Part A):1059–74.
4. Samal P, Vundavilli PR, Meher A, Mahapatra MM. Recent progress in aluminum metal matrix composites: a review on processing, mechanical and wear properties. J Manuf Process. 2020;59:131–52.
5. Kumar PL, Lombardi A, Byczynski G, Murty SVSN, Murty BS, Bichler L. Recent advances in aluminium matrix composites reinforced with graphene-based nanomaterial: a critical review. Prog Mater Sci. 2022;128:100948.
6. Saini DK, Jha PK. Fabrication of aluminum metal matrix composite through continuous casting route: a review and future directions. J Manuf Process. 2023;96:138–60.
7. Parikh VK, Patel V, Pandya DP, Andersson J. Current status on manufacturing routes to produce metal matrix composites: state-of-the-art. Heliyon. 2023;9(2):e13558.
8. Khalid MY, Umer R, Khan KA. Review of recent trends and developments in aluminum 7075 alloy and its metal matrix composites (MMCs) for aircraft applications. Results Eng. 2023;20:101372.
9. Loyd DJ. Particle reinforced aluminum magnesium composites. Int Mater Rev. 1994;39(1): 1–22.
10. Ohring M. Engineering materials science. Academic Press, San Diego; 1995.
11. Askew JR, Wilde JF, Khan TI. TLP bonding of 2124 aluminum metal matrix composite. Mater Sci Technol. 1998;14(5):920–4.
12. Rana RS, Purohit R, Soni VK, Das S. Characterization of mechanical properties and microstructure of aluminum alloy-SiC composites. Mater Today Proc. 2015;2(4–5):1149–56.
13. Shirvanimoghaddam K, Hamim S, Akbari MK, Fakhrhoseini SM, Khayyam H, Pakseresht AH, Ghasali E, Zabet M, Munir KS, Jia SA, Davim JP, Naebe M. Carbon fiber reinforced metal matrix composites: fabrication processes and properties. Compos A: Appl Sci Manuf. 2017;92:70–96.
14. Liao ZR, Abdelhafeez A, Li HN, Yong Y, Diaz OG, Axinte D. State-of-the-art of surface integrity in machining of metal matrix composites. Int J Mach Tools Manuf. 2019;143:63–91.
15. Ramanathan A, Krishnan PK, Muraliraja R. A review on the production of metal matrix composites through stir casting – furnace design, properties, challenges, and research opportunities. J Manuf Process. 2019;42:213–45.
16. Ogawa F, Masuda C. Fabrication and the mechanical and physical properties of nanocarbon-reinforced light metal matrix composites: a review and future directions. Mater Sci Eng A. 2021;820:141542.
17. Fang YJ, Kim MK, Zhang YL, Duan ZY, Yuan Q, Suhr J. Particulate-reinforced iron-based metal matrix composites fabricated by selective laser melting: a systematic review. J Manuf Process. 2022;74:592–639.
18. Maurya P, Kota N, Gibmeier J, Wanner A, Roy S. Review on study of internal load transfer in metal matrix composites using diffraction techniques. Mater Sci Eng A. 2022;840:142973.
19. Isekl T, Kameda T, Maruyama T. Interfacial reaction between SiC and aluminum during joining. J Mater Sci. 1984;19:1692–8.
20. Riedel R. Handbook of ceramic hard materials. Weinheim/New York: Wiley-VCH; 2000.
21. Arik H, Aydin M, Kurt A, Turker M. Weldability of Al_4C_3-Al composites via diffusion welding technique. Mater Des. 2005;26(6):555–60.

22. Schell JSU, Guilleminot J, Binetruy C, Krawczak P. Computational and experimental analysis of fusion bonding in thermoplastic composites: influence of process parameters. J Mater Process Technol. 2009;209(11):5211–9.
23. He TB, Lu TW, Ciftci N, Uhlenwinkel V, Chen WP, Nielsch K, Scudino S. Interfacial characteristics and mechanical asymmetry in Al2024 matrix composites containing Fe-based metallic glass particles. Mater Sci Eng A. 2020;793:139971.
24. Wu S, Qin Y, Fu DR, Fan L, Chen HH, Hong HP. Preparation and mechanism of active Mo-Mn metallization on ZTA particles surface and interfacial bonding of reinforced iron matrix composite. Ceram Int Part B. 2020;46(10):15972–81.
25. Yamamoto M, Nishimura Y, Hayashida M. Influence of Al particles as infiltration promoters on the interfacial reaction and mechanical property of a continuous SiC fiber/AZ91 composite fabricated by a low-pressure infiltration method. J Alloys Compd. 2021;887:161461.
26. Guo SY, Zhang X, Shi CS, Zhao DD, He CN, Zhao NQ. Fully exploiting the role of heterogeneous grain structure in graphene nanoplatelets reinforced Cu matrix composites through interfacial structure design. Compos Commun. 2022;35:101287.
27. Nateq B, Haddad-Sabzevar M, Sajjadi SA, Pellizzari M. Interfacial structure-property relationship in a carbon nanotube-reinforced aluminum alloy matrix composite fabricated by an advanced method. Mater Charact. 2023;203:113159.
28. Bagnato T, Ravindran AR, Mirabedini A, Ladani RB, Kandare E, Orifici AC, Chang P, Wang J, Mouritz AP. Superior interfacial toughening of hybrid metal-composite structural joints using 3D printed pins. Compos A: Appl Sci Manuf. 2023;168:107479.
29. Song MX, Peng K. Pre-oxidization of SiC on interfacial structure and mechanical properties of SiC_p/Al composites prepared by vacuum-pressure infiltration. J Mater Res Technol. 2023;24:6567–77.
30. Everett RK, Arsenault RJ. Metal matrix composites: mechanisms and properties. Boston: Academic Press; 1991.
31. Davis JR. ASM specialty handbook. Aluminum and aluminum alloys. Materials Park: ASM International; 1993. p. 557.
32. Nair SV, Tien JK, Bates RC. SiC-reinforced aluminum metal matrix composites. Int Metals Rev. 1995;30(6):275–88.
33. Gómez de Salazar JM, Barrena MI. Dissimilar fusion welding of AA7020/MMC reinforced with Al_2O_3 particles: microstructure and mechanical properties. Mater Sci Eng A. 2003;352:162–8.
34. Maity J, Pal TK, Maiti R. Transient liquid phase diffusion bonding of 6061-15 wt% SiC_p in argon environment. J Mater Process Technol. 2009;209(7):3568–80.
35. Butt J, Mebrahtu H, Shirvani H. Microstructure and mechanical properties of dissimilar pure copper foil/1050 aluminum composites made with composite metal foil manufacturing. J Mater Process Technol. 2016;238:96–107.
36. Avettand-Fènoël MN, Simar A. A review about friction stir welding of metal matrix composites. Mater Charact. 2016;120:1–17.
37. Chaudhary W, Bharti A, Azam SM, Kumar N, Saxena KK. A re-investigation: effect of TIG welding parameters on microstructure, mechanical, corrosion properties of welded joints. Mater Today Proc. 2021;45(6):4575–80.
38. Hayat F. Electron beam welding of 7075 aluminum alloy: microstructure and fracture properties. Eng Sci Technol Int J. 2022;34:101093.
39. Norouzian M, Elahi MA, Plapper P. A review: suppression of the solidification cracks in the laser welding process by controlling the grain structure and chemical compositions. J Adv Join Process. 2023;7:100139.
40. Hua M, Guo KW, Law HW, Ho JKL. Half – transient liquid phase diffusion welding: an approach for diffusion welding of SiCp/A356 with Cu interlayer. Int J Adv Manuf Technol. 2008;37(5–6):504–12. https://doi.org/10.1007/s00170-007-0983-2.
41. Guo KW. Influence of *in situ* reaction on the microstructure of SiC_p/AlSi7Mg welded by Nd:YAG laser with Ti filler. J Mater Eng Perform. 2009;19:52. https://doi.org/10.1007/s11665-009-9422-x.

42. Okamoto H. Phase diagrams for binary alloys: a desk handbook. Materials Park: ASM International; 2000.
43. Guo W, Hua M, Ho JKL. Study on liquid-phase-impact diffusion welding SiC_p/ZL101. Compos Sci Technol. 2007;67(6):1041–6.
44. Guo W, Hua M, Law HW, Ho JKL. Liquid phase impact diffusion welding of SiC_p/6061Al and its mechanism. Mater Sci Eng A. 2008;490(1–2):427–37.
45. Guo W, Hua M, Ho JKL, Law HW. Mechanism and influence of pulse-impact on properties of liquid phase pulse-impact diffusion welded SiC_p/A356. Int J Adv Manuf Technol. 2009;40(9–10):898–906.
46. Kwok Y. Effects of pulse-impact on the welding of SiC_P/Al-6061 aluminum matrix composites. Mater Sci Technol. 2017;33:2298. https://doi.org/10.1080/02670836.2017.1388604.
47. Meher A, Mahapatra MM, Samal P, Vundavilli PR. A review on manufacturability of magnesium matrix composites: processing, tribology, joining, and machining. CIRP J Manuf Sci Technol. 2022;39:134–58.
48. Aydın F. A review of recent developments in the corrosion performance of aluminum matrix composites. J Alloys Compd. 2023;949:169508.
49. Careri F, Khan RHU, Todd C, Attallah MM. Additive manufacturing of heat exchangers in aerospace applications: a review. Appl Therm Eng. 2023;235:121387.
50. Guo KW. Green nanotechnology of trends in future energy: a review. Int J Energy Res. 2012;36(1):1–17.
51. Heidarzadeh A, Mohammadzadeh R, Brabazon D. Green materials and production of metallic composite materials. Encyclopedia Mater Compos. 2021;2:744–57.
52. Shoji D, He Z, Zhang D, Li VC. The greening of engineered cementitious composites (ECC): a review. Constr Build Mater. 2022;327:126701.
53. Musarurwa H, Tavengwa NT. Smart metal-organic framework (MOF) composites and their applications in environmental remediation. Mater Today Commun. 2022;33:104823.

Chapter 4
Fatigue Failure in Aircraft Materials

Alireza Khalifeh

4.1 Introduction

The primary focus in the aviation sector revolves around ensuring the proper functioning of components and flight safety. Despite considerable efforts directed toward developing materials and enhancing system reliability, instances of premature failures and fatal accidents persist [1–5]. Due to the intricate nature of aviation systems and the interplay among failing components, conducting equipment failure analysis processes is inherently challenging.

Aircraft components frequently endure repeated cyclic loads, which induce cyclic stresses and may result in microscopic damages in the materials [1, 5, 6]. It has been observed that even at stress levels well below the ultimate strength, these microscopic damages can accumulate under the action of cyclic loading until they initiate a crack, ultimately leading to the separation of the component. Moreover, the material inherently contains cracks and other microscopic defects that propagate under cyclic loads, eventually causing fractures in the structural parts. These failures are often sudden, unpredictable, and catastrophic, occurring sometimes after only a short period of service life. Table 4.1 gives a comparison between material failures in general engineering applications and aircraft applications, revealing that the primary source of failures in aircraft is fatigue [7].

Figure 4.1 shows broken studs and their locations in a conventional aircraft engine cylinder. After only 200 hours of flight, inspection revealed that three out of six studs had failed in the engine. The studs were utilized to connect the cylinder base and crankcase. Examination of the fracture surfaces indicated that fatigue was the primary cause of separation and failure of these components. In this instance, fatigue resulted from the movement of the cylinder flange on the crankcase and

A. Khalifeh (✉)
Shiraz Petrochemical Complex, Shiraz, Iran
e-mail: ar.khalifeh@spc.co.ir

S. Gürgen (ed.), *Failure in Aircraft Materials*, Sustainable Aviation,
https://doi.org/10.1007/978-3-031-65850-1_4

Table 4.1 Material failure modes in general engineering and aircraft applications [7, 8]

	Frequency of material failures	
Failure mode	Engineering applications	Aircraft applications
Fatigue	25%	55%
Corrosion	29%	16%
Brittle fracture	16%	–
Overload	11%	14%
High-temperature corrosion	7%	2%
Corrosion fatigue	6%	7%
Wear/erosion	3%	5%
Other	3%	1%

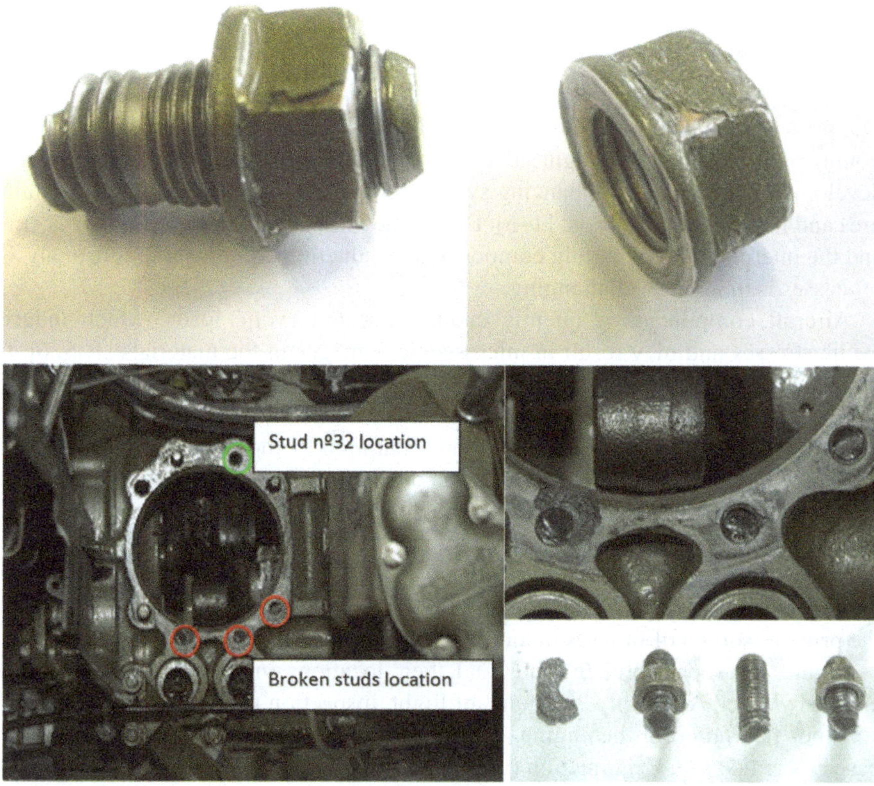

Fig. 4.1 Failed studs and their locations in an aircraft engine [9]

cyclic stress stemming from incorrect bolt preload or its loss [9]. During the preparation for takeoff, an ATR-42 aircraft experienced a collapse of the swinging lever due to fatigue. The alloy involved was manufactured from an AA7010-T74. Fractographs for the broken area are shown in Fig. 4.2. Analysis of the fracture surfaces revealed that fatigue was the root cause of the swinging lever failure. Fatigue initiation occurred on the external surface, corresponding to the most

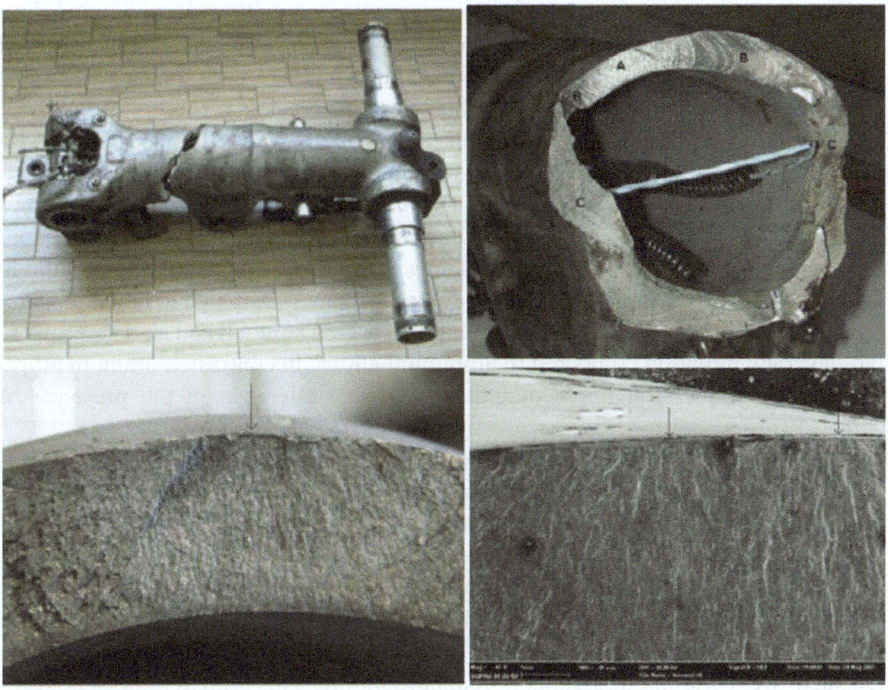

Fig. 4.2 (**a**) General view of swinging lever, (**b**) macroview of fracture surface, (**c**) detailed view of fracture surface, and (**d**) SEM image of fracture surface [5]

stressed area during the forging operation of the alloy. The crack was observed to propagate through approximately 30% of the swinging lever's cross-section, indicating fatigue characteristics [5].

As given in the previous examples, fatigue failures are quite common in the aviation industry. To avoid fatigue failures in aircraft components, and ultimately enhance safety in the sector, material selection comes to the forefront in aircraft design. Understanding material properties such as fracture toughness, fatigue crack initiation, and propagation rates is crucial for preventing fatigue failures. Enhancing fatigue properties also necessitates a thorough understanding of loading conditions, stress states, and material characteristics. The primary aim of this chapter is to present scientific findings and pertinent engineering practices related to fatigue failures in the aviation industry.

4.2 Aviation Materials

Generally, the materials used in the aviation industry are composites, aluminum alloys, titanium alloys, and superalloys [7, 10]. Aluminum alloys are the most widely used materials due to their advanced properties such as high strength-to-density ratio, excellent toughness, and ease of manufacture, including weldability and joinability. AA2024-T3 is one of the most common alloys in aero-structures. For this alloy, the enhancement of fatigue crack growth resistance has been achieved through controlled over-aging [11]. Another aluminum alloy used for aero-structures is AA7075, where the focus lies on multisite fatigue crack nucleation [12].

Composites represent another category of materials widely utilized in the aviation industry, offering advantages such as lighter weight, corrosion resistance, optimized layup, and improved fatigue life. However, the use of composites also comes with some disadvantages, including high material costs, sensitivity to temperature and moisture, susceptibility to impact damage and delamination, weaker strength in the out-of-plane direction, and more challenging repairs compared to metallic materials [7, 13]. Composite materials exhibit high specific strength and are less prone to fatigue crack initiation compared to aluminum alloys. In contrast, aluminum alloys demonstrate higher toughness and better damage tolerance in the presence of defects [7].

Titanium alloys are also widely used in aircraft due to their excellent combination of properties, including high strength-to-weight ratio, corrosion resistance, and high-temperature stability. These alloys are particularly well-suited for critical aircraft components such as airframes, landing gear, engine parts, and structural elements. When it comes to fatigue properties, titanium alloys exhibit several characteristics that make them desirable for aviation applications. Titanium alloys have relatively high fatigue strength, allowing them to withstand cyclic loading and repeated stress without undergoing premature failure. This property is essential for aircraft components subjected to fluctuating loads during flight cycles and ground operations. Furthermore, titanium alloys demonstrate excellent resistance to fatigue crack initiation and propagation. Their microstructure and mechanical properties contribute to the material's ability to resist the growth of fatigue cracks, thereby extending the fatigue life of components [14].

Superalloys are a class of high-performance materials known for their exceptional strength, creep resistance, corrosion resistance, and thermal stability at elevated temperatures. They are mostly used in aircraft engines, gas turbines, and other aerospace applications where extreme operating conditions are encountered. When considering fatigue properties, superalloys exhibit several characteristics that make them well-suited for demanding aerospace environments. These alloys maintain their fatigue resistance even at elevated temperatures, making them suitable for high-temperature applications such as turbine blades, discs, and combustion chambers. These alloys can withstand cyclic thermal and mechanical stresses without significant degradation in fatigue performance.

4.3 Characteristics of Fatigue Failure

Fatigue failures typically occur suddenly and stealthily, without warning. The appearance of fatigue failure is characterized by brittleness and lacks significant deformation. At the macroscopic level, the fatigue failure surface typically aligns

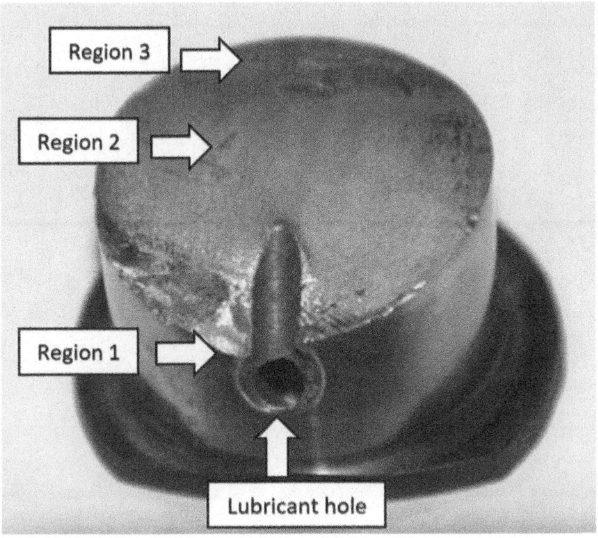

Fig. 4.3 Macroview of a shaft failed by fatigue [15]

perpendicular to the direction of tensile stress. The fracture surface of fatigue failure typically comprises three regions: (i) fatigue crack initiation region, (ii) fatigue expansion or smooth region, and (iii) static fracture region with a rough appearance [15, 16]. An illustration of fatigue failure in a shaft is shown in Fig. 4.3 [15]. The first region is where high stress concentration occurs and the crack initiates. The edge of the lubrication hole of the shaft serves as the site of crack initiation. The second region is the fatigue expansion zone, where the crack initiates and grows. Beach marks, originating from the site of crack initiation and extending into the component, are characteristic of this region. These beach marks, also referred to as progression marks, are crucial for describing fatigue-fractured surfaces. An SEM image of striations is depicted in Fig. 4.4. It should be noted that beach marks are only indicative of fatigue fracture, and some fatigue fractures may not exhibit clear beach marks. The third region is the static fracture region with a rough appearance, where the final catastrophic failure occurs. In this region, the member fails in a ductile manner when the cross-section is unable to withstand the applied load [15, 16].

4.4 Definitions and Concepts

4.4.1 Cyclic Loading

The phenomenon of fatigue occurs due to cyclic loadings that create stress alternations in materials. Figure 4.5 shows various stress conditions during fatigue formation in the materials. A completely reversed cycle of stress in sinusoidal form is

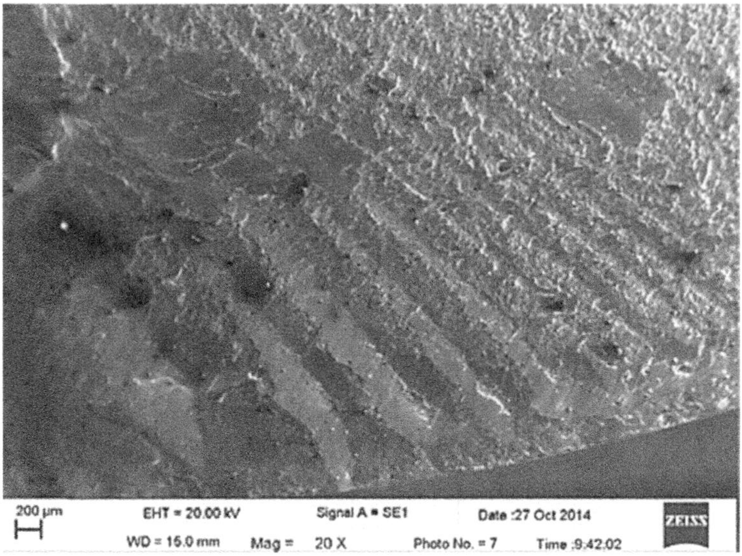

Fig. 4.4 SEM image of beach marks on the fracture surface [15]

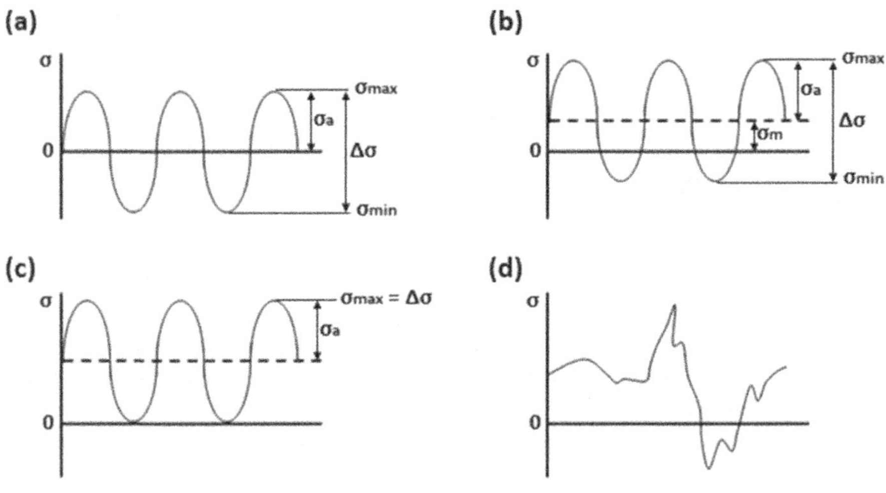

Fig. 4.5 Typical loadings during fatigue: (**a**) completely reversed stress, (**b**) repeated stress or nonzero mean stress, (**c**) zero-to-tensile stress, and (**d**) irregular stress

shown in Fig. 4.5a. This condition is an ideal stress-in-fatigue condition that is achieved in an axially rotating shaft with constant speed and no overload. For this state of cyclic stress, the minimum and maximum stresses are equal to each other. In keeping the conventions, tensile stress is considered positive, compressive stress as negative, and minimum stress means the lowest algebraic stress. Figure 4.5b shows a cyclic stress in which the maximum and minimum stresses are not identical in magnitude. In this case, both stresses can be tensile or compressive, while the maximum and minimum stresses can be in different directions as illustrated in the chart. Another loading is zero-to-tensile stress as shown in Fig. 4.5c. In this case, the minimum stress drops to zero while the maximum stress shows a certain level of loading in the material. Figure 4.5d shows a complicated stress cycle, also known as an irregular stress condition, mainly encountered with unpredictable loads such as those on an airplane wing under the influence of strong winds [16, 17].

Stress range ($\boldsymbol{\sigma}_{\mathbf{r}}$) is described as the difference between the maximum ($\boldsymbol{\sigma}_{\mathbf{max}}$) and minimum ($\boldsymbol{\sigma}_{\mathbf{min}}$) stresses as given in Eq. 4.1.

$$\sigma_{\mathbf{r}} = \sigma_{\max} - \sigma_{\min} \tag{4.1}$$

The half of the range is called stress amplitude ($\boldsymbol{\sigma}_{\mathbf{amp}}$) as given in Eq. 4.2.

$$\sigma_{\mathbf{amp}} = \frac{\sigma_{\max} - \sigma_{\min}}{\mathbf{2}} \tag{4.2}$$

Mean stress ($\boldsymbol{\sigma}_{\mathbf{m}}$) is the average of the maximum and minimum stresses as given in Eq. 4.3.

$$\sigma_{\mathbf{m}} = \frac{\sigma_{\max} + \sigma_{\min}}{\mathbf{2}} \tag{4.3}$$

The stress ratio (**R**) is described as given in Eq. 4.4.

$$\mathbf{R} = \frac{\sigma_{\max}}{\sigma_{\min}} \tag{4.4}$$

4.4.2 Fatigue Life

Fatigue life is mostly characterized by stress–cycle life (S–N) curves. When an engineering material is subjected to cyclic stress, fatigue cracks develop on the surface or subsurface of the material, leading to crack propagation and ultimately material failure. If the material is subjected to higher stresses, the number of cycles to failure gets smaller. Figure 4.6 shows a typical S–N diagram for different materials. Some alloys, such as Material-1 in the chart, show an obvious fatigue limit even as the number of cycles increases. Plain carbon steel is one typical alloy having a fatigue limit under cyclic loading. On the other hand, some materials, such as Material-2 in the chart, do not have a distinct fatigue limit, showing a constant reduction in stress level as the number of cycles increases in the S–N diagram. Aluminum alloys are in this category, having no fatigue limit under cyclic loading.

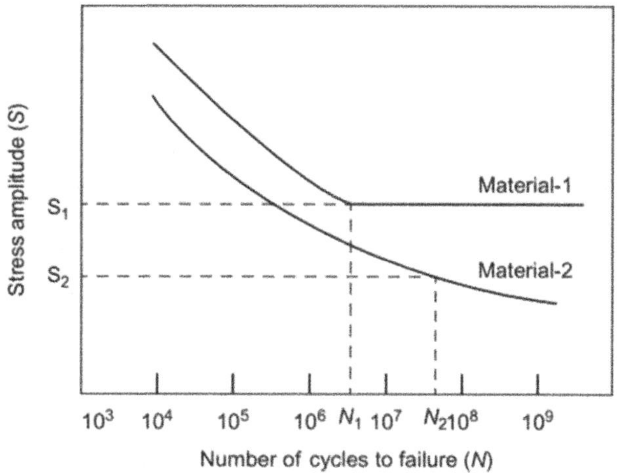

Fig. 4.6 Typical S–N diagram for two different materials [18]

A specific number of cycles is defined to have a fatigue limit for this type of alloys (e.g., 10^8 for aluminum alloys) [16].

Depending on the number of cycles to failure, fatigue can be classified into three different groups: (i) high cycle fatigue, (ii) low cycle fatigue, and (iii) extremely low cycle fatigue. High cycle fatigue is a state of fatigue in which the component is subjected to a high number of cycles, generally greater than 10^5, before failure. The failure is caused by small magnitudes of stresses, yielding elastic strains in the material [17]. Stresses in low cycle fatigue are sufficiently high to produce plastic deformations in the material. Due to the higher stress levels in the fatigue process, a lower number of cycles is mostly sufficient for failure of materials. For this reason, the number of cycles is generally lower than 10^5 in low cycle fatigue. Low cycle fatigue is often observed in structures that suffer from thermal loads, creating cyclic stresses [19]. Extremely low cycle fatigue is characterized by large plastic strains in each cycle. In this fatigue mode, strains are of the order of several times the yield strains, which produce very low fatigue life, generally lower than 10^2 [20–23]. Recent studies show that failure of materials under extremely low cycle fatigue is mainly caused by ductile damage [24, 25].

4.4.3 Fatigue Crack Growth

The presence of cracks can significantly compromise the strength of engineering materials. Cracks may exist in components such as bridges, ship structures, pressure vessels, metal structures in large aircraft, and large forgings like turbine rotors, either inherently or developed during service. However, through meticulous inspection programs during manufacturing, it is rare to find cracks or large defects of critical size inside an object. In more common scenarios, small defects within the parts propagate and evolve into cracks under cyclic loading conditions. As these cracks reach the critical size, they precipitate brittle failures in the components. The process of crack propagation under cyclic loading is termed fatigue crack growth [17].

Considerable research has been devoted to determining the crack growth law, aiming to establish a reliable framework that indicates the safe mode of loading for operating components without risking catastrophic failure. For a specific material, the crack growth behavior is typically described by the relationship between the cyclic crack growth rate (da/dN) and the stress intensity factor range (ΔK). An illustrative example of test data and a fitted curve for a material, plotted on a log–log scale, is presented in Fig. 4.7. Notably, in the intermediate region of the graph, the data exhibit a linear pattern. This linear relationship is expressed by Paris' law [17] as given in Eq. 4.5.

$$\frac{da}{dN} = C\left(\Delta K\right)^{m} \tag{4.5}$$

where C is a constant and m is the slope on the log–log plot.

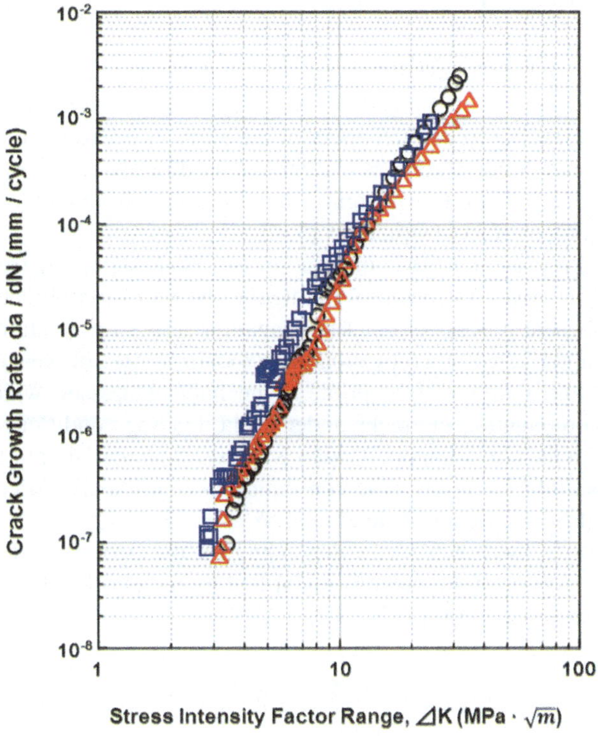

Fig. 4.7 Fatigue crack growth rate of Ti-15Zr-4Nb-4Ta alloy [26]

K is defined as the stress intensity factor, depending on the combination of crack length (a), applied stress (σ), and geometry factor (F), as given in Eq. 4.6.

$$K = F\sigma\sqrt{\pi a} \tag{4.6}$$

The stress intensity factor range (ΔK) can be calculated by using Eq. 4.7.

$$\Delta K = K_{\max} - K_{\min} = F\Delta\sigma\sqrt{\pi a} \tag{4.7}$$

where the stress range can be found as $\Delta\sigma = \sigma_{\max} - \sigma_{\min}$.

At low crack growth rates, the curve typically steepens and appears to converge towards a vertical asymptote represented as K_{th}, termed the fatigue crack growth threshold. This value signifies the lower limit of K below which crack growth typically does not occur. Conversely, at high growth rates, the curve may steepen once again, indicative of rapid and unstable crack growth [17].

4.5 Factors Affecting Fatigue Properties

4.5.1 Notch Effect

Discontinuities such as holes, fillets, grooves, and keyways are unavoidable in the design and construction of engineering parts. These discontinuities result in stress accumulations, leading to a notch effect in the parts. It is essential to consider the notch effect in structural design, as it significantly diminishes fatigue resistance in structures. In fact, a majority of fatigue cracks originate from stress accumulations caused by the notch effect in structures [5, 6].

Figure 4.8 shows a notch effect in a compressor blade stemming from a corrosion pit at the leading edge. As shown in the figure, the corrosion pit acts as the crack origin, initiating the process of crack propagation across the fracture surface. The pit leads to a notch effect, thereby creating a localized area of weakened material. This weakened area is more susceptible to mechanical stresses, which lead to the initiation and propagation of cracks under cyclic loading conditions. As the crack propagates from the corrosion pit, it extends across the material until failure occurs, potentially resulting in catastrophic consequences, particularly in critical components such as the compressor blade of aircraft.

The severity of the notch varies depending on the geometric design [28–31]. Typically, the severity of the notch is quantified by the elastic stress concentration factor (K_t), as expressed in Eq. 4.8.

$$K_t = \sigma / S \tag{4.8}$$

where σ is the local stress at notch and S is the nominal stress.

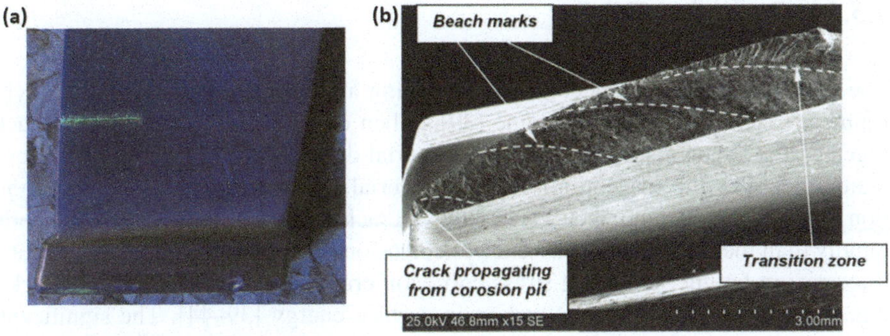

Fig. 4.8 Fatigue failure due to a corrosion pit: (**a**) crack growth under UV light and (**b**) crack propagation stages on fracture surface [27]

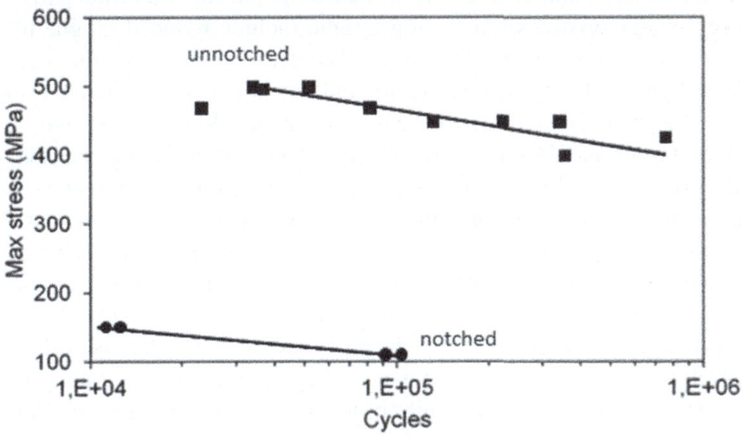

Fig. 4.9 Notch effect on the fatigue behavior of AA7075-RRA [38]

The stress concentration factor can be determined using elasticity theory, and stress concentration factors for various geometries have been compiled by Patterson [32]. The impact of the notch on the fatigue strength of materials is evident when comparing the S–N curves of samples with and without a notch. Figure 4.9 shows the S–N curves of an unnotched and a notched AA7075 under retrogression and re-aging (RRA) conditions. As depicted in the chart, fatigue strength significantly decreases in the presence of a notch. The reduction in fatigue limit is quantified by the fatigue strength reduction factor or the fatigue notch factor. This factor is commonly defined as the ratio of the fatigue strength of a notched specimen to the fatigue strength of a similar specimen without the notch [33]. It is important to note that the fatigue notch factor depends on various conditions such as notch type, notch severity, material, type of loading, stress level, etc. [16, 34–37].

4.5.2 Corrosion Effect

The simultaneous occurrence of fatigue loading and exposure to a corrosive environment is termed corrosion fatigue [39]. When corrosion and cyclic loading act together, they synergistically accelerate material degradation. The presence of corrosion increases the susceptibility of the material to crack initiation and propagation. The cyclic loading then causes these cracks to grow and propagate more rapidly than they would in the absence of corrosion. Several mechanisms contribute to corrosion fatigue, including stress corrosion cracking, hydrogen-assisted cracking, anodic dissolution, and reduction in surface energy [39–44]. The significant effects of corrosion fatigue include surface irregularities, thickness reduction, surface roughness, and stress concentration resulting from pitting and material degradation due to electrochemical corrosive processes. For instance, corrosion pits on component surfaces, similar to notches, reduce the fatigue life of the parts.

The key characteristics of corrosion fatigue include reduced fatigue life, quick crack propagation, and localized damage. The presence of corrosion significantly reduces the fatigue life of a material compared to when it is subjected to cyclic loading alone. The presence of a corrosive environment accelerates the growth of fatigue cracks in materials and diminishes their fatigue life. Research indicates that materials exhibiting a fatigue limit in air and room temperature fatigue tests may not demonstrate a fatigue limit when subjected to a corrosive environment [45]. A comparison of fatigue life for two different aluminum alloys is presented in Fig. 4.10. As depicted in the graph, the fatigue strength of the alloys decreases in the presence of a corrosive environment. Furthermore, corrosion accelerates the initiation and propagation of cracks within the parts due to the brittleness of the material. These cracks can grow under the influence of cyclic loading, eventually leading to failure. Corrosion fatigue often results in localized damage, with cracks initiating and propagating from areas where corrosion is most severe.

4.5.3 Thermal Effect

Thermal fatigue in aviation materials refers to the phenomenon of cyclic thermal stresses leading to material degradation and eventual failure in aircraft components. It is a critical consideration in aerospace engineering due to the extreme temperature variations experienced during service. Aircraft materials are subjected to wide temperature ranges during different phases of flight, from low temperatures at high altitudes to high temperatures during takeoff and landing. These rapid and extreme temperature changes induce thermal expansion and contraction in various components, generating internal stresses within the materials [47].

Different materials used in aviation, such as metals, composites, and coatings, have different coefficients of thermal expansion and mechanical properties. As temperatures fluctuate, these materials expand and contract at different rates, leading to

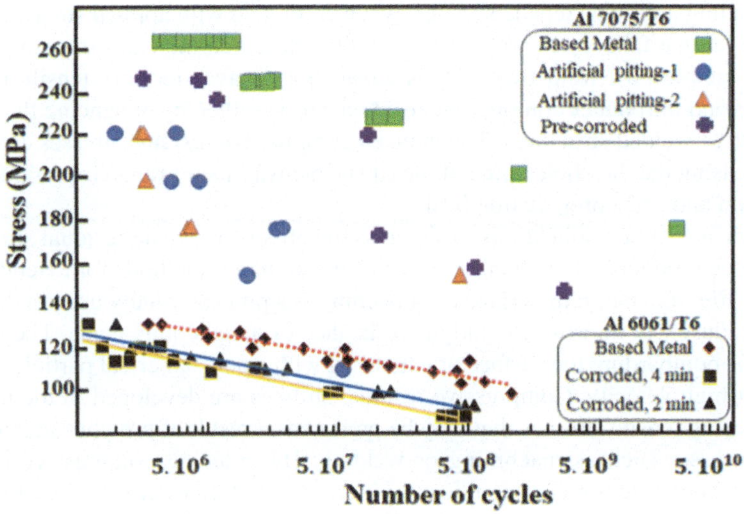

Fig. 4.10 Effect of corrosion on fatigue life [46]

the development of thermal stresses within the components. Moreover, thermal gradients within the materials can create stress concentrations at points of geometric irregularities, such as sharp corners, edges, fastener holes, and material interfaces. These stress concentration points become susceptible to crack initiation under thermal cycling conditions. As the aircraft undergoes repeated flight cycles, the crack initiations turn into microcracks, propagating, growing, and eventually leading to macroscopic cracks and component failure.

4.5.4 Residual Stress Effect

Residual stresses arise in a part due to manufacturing and fabrication processes, where external loads are often a significant proportion of the yield strength [23, 48]. These stresses exert a considerable influence on the fatigue performance of engineering components. Depending on the alignment of residual stresses relative to applied stresses, they can either enhance or diminish the fatigue resistance of the components [49–55]. However, the substantial residual stresses generated during manufacturing processes are often inadequately considered by designers, potentially leading to premature failures of components [56]. The presence of residual stresses in a component affects both the initiation and propagation stages of fatigue cracks due to alterations in the effective mean stress experienced during fatigue cycling [51].

Welding, as a joining process for manufacturing parts, involves heating cycles that induce significant changes in material microstructure and result in the generation of high-magnitude residual stresses in the weldments [57]. Throughout the welding process, residual compressive and tensile stresses form in the weld and the surrounding area. The tensile stresses, when combined with applied stress, can hasten the fatigue failure of components [57–60]. Research indicates that implementing a suitable post-weld treatment is advantageous as it reduces tensile residual stresses and introduces compressive residual stresses, thereby enhancing the fatigue strength of welded structures. The induced compressive residual stresses contribute to increasing the threshold value of the stress intensity factor range for fatigue crack initiation and early propagation [60].

There are some applications to suppress the effects of tensile residual stresses in aerospace products. Shot peening is one of the common methods for extending the fatigue life of aviation materials. Shot peening is a process widely used in the aerospace industry to improve the fatigue resistance of metal components. The process involves bombarding the surface of a material with small, spherical particles, called shot, at high velocity. Compressive residual stresses are developed in the material by relieving tensile stresses that may be present as a result of previous manufacturing processes such as machining or welding. By inducing compressive residual stresses, shot peening can significantly increase the fatigue life of metal components, making them less prone to cracking and failure under cyclic loading conditions. Furthermore, the impact of the shot can work-harden the surface of the material, increasing its resistance to wear and abrasion.

4.6 Conclusion

Fatigue failure of aircraft materials is of paramount importance due to the significant risk it poses to human life. Aircraft components subjected to cyclic loading over time are prone to fatigue failure, which can lead to catastrophic consequences such as structural failure or in-flight accidents. Therefore, ensuring the reliability and structural integrity of aircraft materials is crucial for aviation safety and the protection of human lives. Vigilant inspection, maintenance, and adherence to rigorous safety standards are essential measures to mitigate the risks associated with fatigue failure in aircraft materials.

Depending on the material and loading conditions, fatigue failure can manifest as high cycle, low cycle, or extremely low cycle fatigue. Materials utilized in the aerospace industry must exhibit specific characteristics, including a certain resistance to fatigue failure. The presence of harsh environments can induce corrosion fatigue, leading to a further reduction in the fatigue life of components. Similarly, thermal cycles can facilitate crack initiation and propagation, thereby reducing the fatigue resistance of materials. On the other hand, residual stresses originating from the manufacturing process of parts are superimposed onto the stresses developed during service, inducing the fatigue failure of components.

Understanding the mechanisms of fatigue, implementing rigorous inspection and maintenance practices, and employing robust design and materials solutions are essential for mitigating the risks associated with fatigue-related failures in aircraft structures.

References

1. Lv J, Wang W, Liu S. Statistical analysis of failure cases in aerospace. Int J Aerosp Mech Eng. 2018;12(5):497–501.
2. Sun Z, et al. Failure mechanism and reliability analysis of the rear cargo door lock of transport aircraft. Eng Fail Anal. 2021;122:105182.
3. Freitas M, Infante V, Baptista R. Failure analysis of the nose landing gear axle of an aircraft. Eng Fail Anal. 2019;101:113–20.
4. Guo W, et al. Function failure and failure boundary analysis for an aircraft lock mechanism. Eng Fail Anal. 2016;70:428–42.
5. Bagnoli F, et al. Fatigue fracture of a main landing gear swinging lever in a civil aircraft. Eng Fail Anal. 2008;15(6):755–65.
6. Pang H, Yu T, Song B. Failure mechanism analysis and reliability assessment of an aircraft slat. Eng Fail Anal. 2016;60:261–79.
7. Tavares S, De Castro P. An overview of fatigue in aircraft structures. Fatigue Fract Eng Mater Struct. 2017;40(10):1510–29.
8. Findlay S, Harrison N. Why aircraft fail. Mater Today. 2002;5(11):18–25.
9. Duarte D, et al. An overview on how failure analysis contributes to flight safety in the Portuguese Air Force. Eng Fail Anal. 2016;65:86–101.
10. Doré MJ, Maddox SJ. Accelerated fatigue crack growth in 6082 T651 aluminium alloy subjected to periodic underloads. Proc Eng. 2013;66:313–22.
11. Tzamtzis A, Kermanidis A. Improvement of fatigue crack growth resistance by controlled overaging in 2024-T3 aluminium alloy. Fatigue Fract Eng Mater Struct. 2014;37(7):751–63.
12. Jin Y, et al. An experimental methodology for quantitative characterization of multisite fatigue crack nucleation in high-strength Al alloys. Fatigue Fract Eng Mater Struct. 2016;39(6):696–711.
13. Campbell FC. Structural composite materials. Materials Park: ASM International; 2010.
14. Bache MR. A review of dwell sensitive fatigue in titanium alloys: the role of microstructure, texture and operating conditions. Int J Fatigue. 2003;25:1079–87.
15. Li W, Yan Q, Xue J. Analysis of a crankshaft fatigue failure. Eng Fail Anal. 2015;55:139–47.
16. Dieter GE, Bacon D. Mechanical metallurgy, vol. 3. New York: McGraw-Hill; 1976.
17. Dowling N. Mechanical behaviour of materials. Engineering methods for deformation, fracture, and fatigue. 4th ed. Harlow: Pearson Education Limited; 2013.
18. Gürgen S, Kuşhan MC, Diltemiz SF. Fatigue failure in aircraft structural components. In: Handbook of materials failure analysis with case studies from the aerospace and automotive industries. Oxford: Elsevier; 2016.
19. Manson SS, Dolan T. Thermal stress and low cycle fatigue. J Appl Mech. 1966;33(4):957.
20. Yang L, et al. Low cycle fatigue property and fracture behavior of low yield point steels. Constr Build Mater. 2018;165:688–96.
21. Tateishi K, Hanji T, Minami K. A prediction model for extremely low cycle fatigue strength of structural steel. Int J Fatigue. 2007;29(5):887–96.
22. Nip K, et al. Extremely low cycle fatigue tests on structural carbon steel and stainless steel. J Constr Steel Res. 2010;66(1):96–110.
23. Khalifeh A. Perspective chapter: fatigue of materials. In: Failure analysis-structural health monitoring of structure and infrastructure components. London: IntechOpen; 2023.
24. Xue L. A unified expression for low cycle fatigue and extremely low cycle fatigue and its implication for monotonic loading. Int J Fatigue. 2008;30(10–11):1691–8.
25. Kuroda M. Extremely low cycle fatigue life prediction based on a new cumulative fatigue damage model. Int J Fatigue. 2002;24(6):699–703.
26. Okazaki Y. Comparison of fatigue properties and fatigue crack growth rates of various implantable metals. Materials. 2012;5(12):2981–3005.
27. Witek L. Crack propagation analysis of mechanically damaged compressor blades subjected to high cycle fatigue. Eng Fail Anal. 2011;18(4):1223–32.

28. Saini DS, Karmakar D, Ray-Chaudhuri S. A review of stress concentration factors in tubular and non-tubular joints for design of offshore installations. J Ocean Eng Sci. 2016;1(3):186–202.
29. Hellier A, Connolly M, Dover W. Stress concentration factors for tubular Y-and T-joints. Int J Fatigue. 1990;12(1):13–23.
30. Ozkan MT, Erdemir F. Determination of stress concentration factors for shafts under tension. Mater Testing. 2020;62(4):413–21.
31. Santos A, et al. Simulation of stress concentration factors in combined discontinuities on flat plates. In: Journal of physics: conference series. IOP Publishing; 2016.
32. Pilkey WD, Pilkey DF, Bi Z. Peterson's stress concentration factors. Hoboken: Wiley; 2020.
33. Algarni M. Notch factor correction using stress triaxiality of plane-stress state in high-cycle fatigue. Int J Fatigue. 2019;128:105204.
34. Yao W, Xia K, Gu Y. On the fatigue notch factor. Kf International Journal of Fatigue. 1995;17(4):245–51.
35. Hu Z-Z, Cao S-Z. Relationship between fatigue notch factor and strength. Eng Fract Mech. 1994;48(1):127–36.
36. Owolabi G, Prasannavenkatesan R, McDowell D. Probabilistic framework for a microstructure-sensitive fatigue notch factor. Int J Fatigue. 2010;32(8):1378–88.
37. McEvily A, et al. Fatigue notch sensitivity and the notch size effect. Int J Fatigue. 2008;30(12):2087–93.
38. Camarinha MGG, et al. Fatigue behavior of notched and unnotched 7075-T6 aluminum alloy subjected to retrogression and re-aging (RRA) heat treatment and plasma nitriding. Theor Appl Fract Mech. 2023;127:104051.
39. Cracking H. Mechanisms of corrosion fatigue. Fatigue and fracture. 1996;19:185–92.
40. Parkins R, Greenwell B. The interface between corrosion fatigue and stress-corrosion cracking. Metal Science. 1977;11(8–9):405–13.
41. Wang Y. Corrosion fatigue. In: Uhlig's corrosion handbook, vol. 3. Hoboken: Wiley; 2011. p. 195–202.
42. Dolley L, Wei. The effect of pitting corrosion on fatigue life. Fatigue Fract Eng Mater Struct. 2000;23(7):555–60.
43. Pao P, Feng C, Gill S. Corrosion fatigue crack initiation in aluminum alloys 7075 and 7050. Corrosion. 2000;56(10):1022–31.
44. Chlistovsky R, Heffernan P, DuQuesnay D. Corrosion-fatigue behaviour of 7075-T651 aluminum alloy subjected to periodic overloads. Int J Fatigue. 2007;29(9–11):1941–9.
45. Adasooriya N, Hemmingsen T, Pavlou D. Fatigue strength degradation of metals in corrosive environments. In: IOP conference series: materials science and engineering. Stavanger: IOP Publishing; 2017.
46. Behvar A, Haghshenas M. A critical review on very high cycle corrosion fatigue: mechanisms, methods, materials, and models. J Space Saf Eng. 2023;10(3):284–323.
47. Sun Y, et al. A mini-review on the thermal fatigue properties of copper materials applied at the front-end of synchrotron radiation facilities. Entropy. 2023;25(5):714.
48. Khalifeh A. Stress corrosion cracking damages. In: Failure analysis. London: IntechOpen; 2019.
49. James MN, et al. Residual stresses in condition monitoring and repair of thermal power generation components. Theor Appl Fract Mech. 2017;92:289–97.
50. Wang Y, et al. Gradient speed control method to reduce the residual stress on a turbine disk in forging process. In Advances in materials processing: proceedings of Chinese materials conference 2017 18th. Springer; 2018.
51. James M, et al. Residual stresses and fatigue performance. Eng Fail Anal. 2007;14(2):384–95.
52. De Giorgi M. Residual stress evolution in cold-rolled steels. Int J Fatigue. 2011;33(3):507–12.
53. Gothivarekar S, et al. Effect of bending process on the fatigue behaviour of high strength steel. J Constr Steel Res. 2021;182:106662.
54. Han S, Lee T, Shin B. Residual stress relaxation of welded steel components under cyclic load. Steel Res. 2002;73(9):414–20.

55. Quesnel D, Meshii M, Cohen J. Residual stresses in high strength low alloy steel during low cycle fatigue. Mater Sci Eng. 1978;36(2):207–15.
56. Fairfax E, Steinzig M. A summary of failures caused by residual stresses. In: Residual stress, thermomechanics & infrared imaging, hybrid techniques and inverse problems, volume 9: proceedings of the 2015 annual conference on experimental and applied mechanics. New York: Springer; 2016.
57. Manai A, Al-Emrani M. A probabilistic study of welding residual stresses distribution and their contribution to the fatigue life. Eng Fail Anal. 2020;118:104787.
58. Lu X. Influence of residual stress on fatigue failure of welded joints. North Carolina: North Carolina State University; 2003.
59. Gadallah R, et al. Critical investigation on the influence of welding heat input and welding residual stress on stress intensity factor and fatigue crack propagation. Eng Fail Anal. 2018;89:200–21.
60. Cheng X, et al. Residual stress modification by post-weld treatment and its beneficial effect on fatigue strength of welded structures. Int J Fatigue. 2003;25(9–11):1259–69.

Chapter 5
High-Cycle Fatigue Fracture and Corrosion Behavior of an Aerospace Alloy Manufactured with the Wire-Arc Directed Energy Deposition Technology

Xinpeng Guo, Huijun Li, Zengxi Pan, Bosheng Dong, Zhijun Qiu, Chengxun Zhang, and Hu Huang

5.1 Introduction

Aluminum alloys of the 7xxx series, specifically Al-Zn-Mg-Cu compositions, find extensive application in the aerospace industry. They are favored for their exceptional strength-to-weight ratio, good corrosion resistance, and ease of shaping [1–3]. For the present, the primary approach for manufacturing 7xxx series aluminum alloys involves multiple operations including casting and forging, resulting in lengthy cycles of production and suboptimal material utilization. To improve production efficiency and the buy-to-fly ratio, researchers have turned their attention to wire-arc directed energy deposition (WA-DED) technology. This method offers advantages like high material utilization and structural flexibility, which also proves

X. Guo · H. Li (✉) · Z. Pan · C. Zhang
School of Mechanical, Materials, and Mechatronics and Biomedical Engineering, University of Wollongong, Wollongong, NSW, Australia
e-mail: huijun@uow.edu.au

B. Dong
School of Mechanical, Materials, and Mechatronics and Biomedical Engineering, University of Wollongong, Wollongong, NSW, Australia

School of Aerospace, Mechanical and Mechatronic Engineering, University of Sydney, Sydney, NSW, Australia

Z. Qiu
School of Mechanical, Materials, and Mechatronics and Biomedical Engineering, University of Wollongong, Wollongong, NSW, Australia

Australian Nuclear Science and Technology Organization (ANSTO), Lucas Heights, NSW, Australia

H. Huang
Department of Civil Engineering, Monash University, Clayton, VIC, Australia

S. Gürgen (ed.), *Failure in Aircraft Materials*, Sustainable Aviation,
https://doi.org/10.1007/978-3-031-65850-1_5

that WA-DED is a viable and efficient approach for the production of 7xxx series aluminum alloys [4–8]. However, 7xxx aluminum alloys manufactured by WA-DED face two pressing issues. The first one is the problem of element segregation caused by rapid cooling rates, leading to the formation of a significant amount of brittle eutectic phases along grain boundaries, which severely affects the corrosion resistance of the material. 7xxx aluminum alloys manufactured by traditional processes often employ post-heat treatments to address the issue of compositional segregation [9]. Therefore, the segregation problem arising in the WA-DED process can also be effectively addressed through various subsequent heat treatment procedures. Chen et al. investigated the impact of T6, T74, and retrogression and re-aging (RRA) heat treatments on the mechanical properties and susceptibility to exfoliation corrosion of 7085 aluminum alloy. According to the results, the corrosion resistance in increasing order was T6, RRA, and T74. However, the ultimate tensile point in the T74 condition was the lowest, which was about 9.6% lower compared to that achieved with the T6 condition. Moreover, the yield strength (YS) and ultimate tensile strength (UTS) of the RRA heat-treated sample exhibited a slight decrease in comparison to the sample with the T6 condition [10]. Li et al. conducted similar research on the influence of T6, T73, and RRA conditions on the mechanical behavior and intergranular corrosion properties of 7075 aluminum alloy. The corrosion resistance trend closely paralleled the findings of the aforementioned study [11]. The aforementioned research clearly shows that the mechanical properties and corrosion resistance of 7xxx aluminum alloys undergo significant changes with different heat treatment treatments. In general, the T6 heat treatment often results in higher strength but lower corrosion resistance compared to overaging treatments such as T74 or T73. Conversely, overaging treatments typically lead to a certain loss of strength. To balance corrosion resistance and mechanical performance, RRA treatment is often employed to ensure overall material performance.

The second serious problem in WA-DED-manufactured aluminum alloys is the porosity defect. This problem arises from the major difference in hydrogen solubility between the solid and liquid phases, leading to pore formation based on hydrogen gases in the solidification process [12]. In the case of 7xxx aluminum alloys, in addition to pores based on hydrogen gases, the presence of zinc elements can lead to the evaporation of the zinc contents upon arc exposure, resulting in the formation of zinc gas pores as well, further exacerbating the tendency for porosity [8]. Porosity defects can have a particularly detrimental impact on the mechanical properties, especially the fatigue performance of additive-manufactured aluminum alloys, which often become the initial site of fatigue fractures [13].

This chapter presents a detailed study on these two major issues in the WA-DED of 7xxx aluminum alloys. The corrosion behaviors of WA-DED-manufactured 7xxx aluminum alloys were investigated under three different heat treatment processes. Additionally, the chapter investigates the fatigue performance and fracture mechanisms in WA-DED 7xxx aluminum alloys.

5.2 Experimental Details

To investigate the corrosion behaviors of WA-DED 7xxx aluminum alloy under different heat treatments, the self-prepared Al-Zn-Mg-Cu filler wire, namely 7B55-Sc, was used as the feedstock, and the cold metal transfer (CMT) process was employed to prepare thin-walled structures as shown in Fig. 5.1a, b. The chemical compositions of the feedstock and printing parameters are given in Tables 5.1 and 5.2, respectively. Three different heat treatment processes, namely T6, T73, and RRA, were applied to the specimens. The specific heat treatment process steps can be seen in Fig. 5.1c. Electrochemical corrosion testing was performed using a CHI660E-type electrochemical system to investigate the corresponding corrosion resistance. The surface morphologies of the samples were also analyzed to understand the corrosion process on the materials. For fatigue testing, an Instron E20000 fatigue testing machine was employed, operating at a load frequency of 50 Hz and a stress ratio of 0.1. The tests were stopped either upon specimen fracture or after reaching

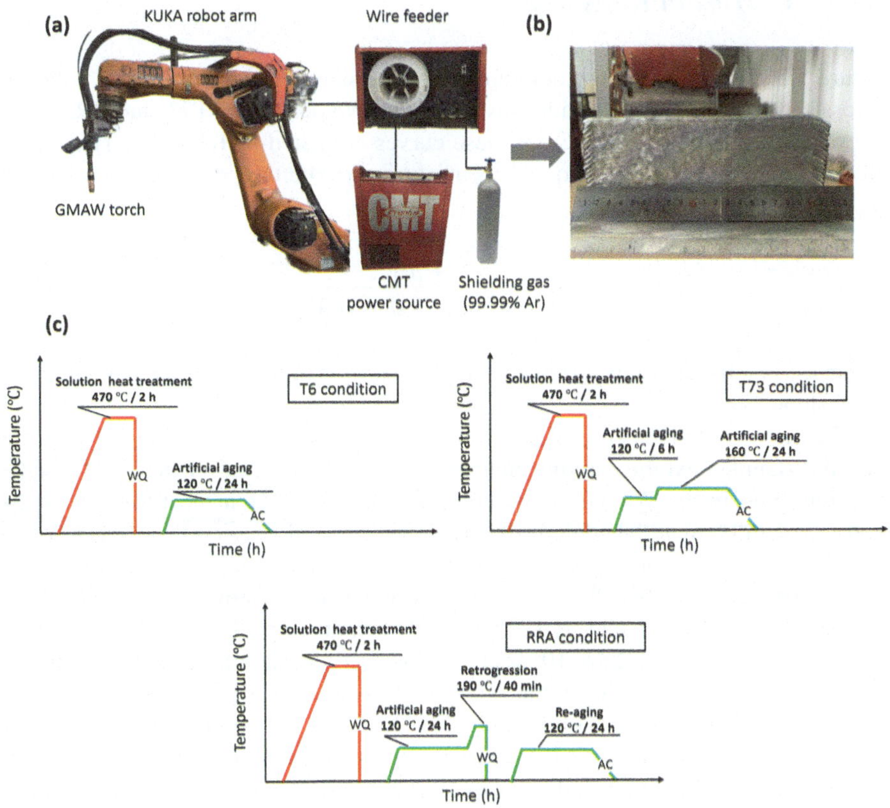

Fig. 5.1 (**a**) WA-DED setup, (**b**) thin-walled component, (**c**) schematic diagrams of heat treatments [9]

Table 5.1 Chemical compositions of the 7B55-Sc wire (wt.%)

Samples	Zn	Mg	Cu	Sc	Zr	Fe	Si	Al
Wire component	7.80	2.12	1.76	0.24	0.18	0.041	0.028	Bal.

Table 5.2 Printing parameters for thin-walled components under CMT mode

Wire feed speed (m/min)	Current (A)	Voltage (V)	Deposition speed (mm/s)
6.5	112	12.9	9

1×10^7 fatigue cycles. S–N curves were generated through linear fitting based on the experimental data. The observation of fatigue fracture morphologies was also performed to analyze the fracture mechanisms.

5.3 Results and Discussion

5.3.1 Corrosion Resistance

Figure 5.2 shows the polarization curves for the samples in a 3.5% NaCl solution across various heat-treated conditions, including WA-DED, T6, T73, and RRA. The relevant parameters obtained from these curves are listed in Table 5.3. The Stern-Geary equation, as given in Eq. 5.1, is used to calculate the polarization resistance (R_p) in the samples [14].

$$R_p = \frac{\beta_a \beta_c}{2.303 i_{corr} (\beta_a + \beta_c)} \tag{5.1}$$

where i_{corr} is self-corrosion current density, β_a is anodic Tafel slope, and β_c is cathodic Tafel slope.

The findings derived from Fig. 5.2 and the data provided in Table 5.3 indicate that the samples exhibit varying magnitudes of self-passivation in a 3.5% NaCl solution, both in the WA-DED condition and across different heat-treated conditions. Generally, materials demonstrating lower i_{corr} and i_p, as well as larger ΔE_{pass} and E_{pit} values, exhibit enhanced self-passivation capabilities and superior corrosion resistance [15, 16]. The T73 heat-treated sample shows the smallest i_{corr} (12.50×10^{-6} A/cm^2) and i_p (2.495×10^{-5} A/cm^2) while the highest ΔE_{pass} (487.95 mV) and R_p (7.228×10^4 Ω.cm^2). However, the samples after T6 treatment exhibit an entirely opposite trend, whereas the values of the samples after RRA treatment fall between those of T6 and T73. Therefore, the observed patterns for all these parameters indicate that the corrosion resistance ranking for the samples, in the order of as-deposited and various heat treatment conditions, is as follows: WA-DED < T6 < RRA < T73.

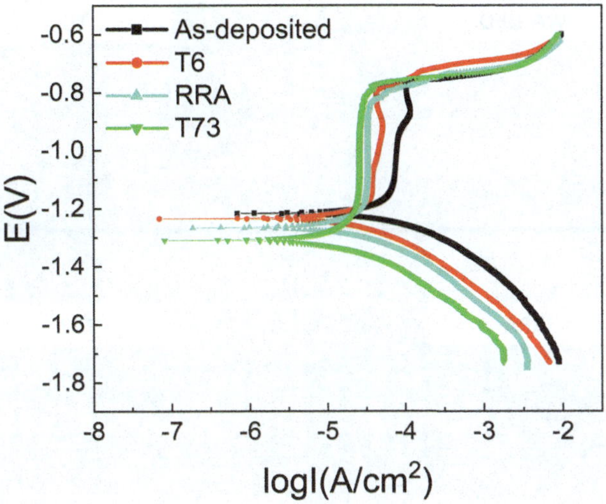

Fig. 5.2 Polarization curves for the samples across different heat treatments [9]

Table 5.3 Fitting parameters of polarization curves for the samples across different heat treatments [9]

Conditions	i_{corr} (A/cm^2)	E_{corr} (V)	R_p (Ω.cm^2)	i_p (A/cm^2)	E_{pit} (V)	ΔE_{pass} (mV)
WA-DED	51.64×10^{-6}	−1.216	1.369×10^4	9.772×10^{-5}	−0.7747	339.40
T6	29.04×10^{-6}	−1.230	2.260×10^4	4.592×10^{-5}	−0.7938	380.61
RRA	20.75×10^{-6}	−1.267	3.433×10^4	3.264×10^{-5}	−0.8236	406.54
T73	12.50×10^{-6}	−1.308	7.228×10^4	2.495×10^{-5}	−0.7678	487.95

5.3.2 *Surface Corrosion Morphologies*

The surface corrosion morphologies across different treated conditions are shown in Fig. 5.3. It is clear that the WA-DED sample exhibits the most pronounced intergranular corrosion, characterized by the formation of continuous corrosion paths along the grain boundaries and the presence of huge corrosion pits as shown in Fig. 5.3a. This is directly due to the continuous $\eta_{Mg(Zn,Cu,Al)_2}$ grain boundaries precipitates (GBPs). As shown in Fig. 5.4a, these continuous GBPs readily give rise to broad and uninterrupted corrosion paths under electrochemical corrosion conditions. Consequently, the WA-DED samples exhibit the lowest corrosion resistance. In the case of the T6 conditioned sample, there is still evident intergranular corrosion, with numerous fine corrosion pits appearing in localized areas along the grain boundaries. However, the extent of corrosion is significantly reduced compared to the WA-DED condition as shown in Fig. 5.3b. From this fact, the root cause is that after the T6 heat treatment, fine precipitates are reprecipitated along the grain

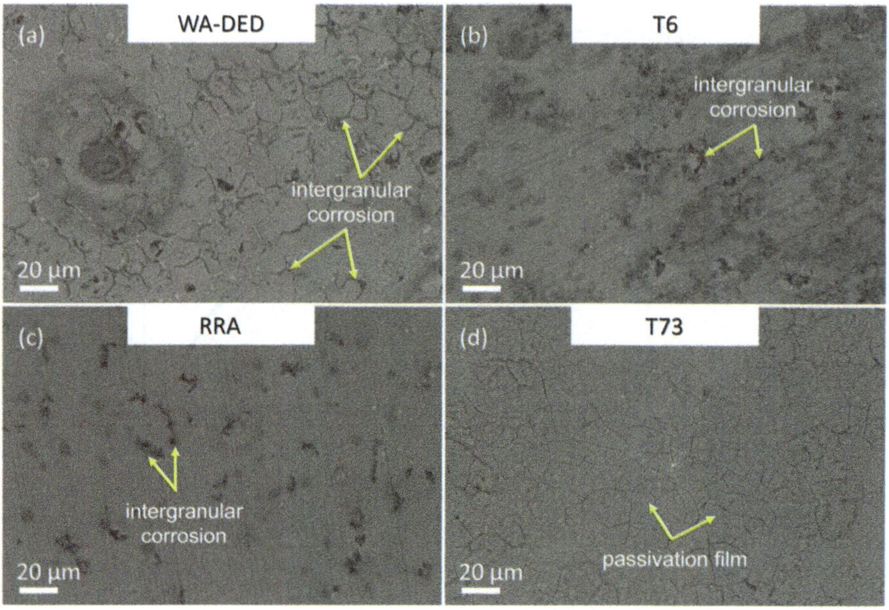

Fig. 5.3 Surface corrosion morphologies of conditions: (**a**) WA-DED, (**b**) T6, (**c**) T73, (**d**) RRA [9]

boundaries compared to the WA-DED condition. However, they still exist in a continuous distribution along the grain boundaries as shown in Fig. 5.4b, hence exhibiting relatively poor corrosion resistance. In the case of the RRA-conditioned sample, intergranular corrosion is observed merely at selective grain boundaries, suggesting an improved level of corrosion resistance compared to the T6 condition as shown in Fig. 5.3c. As we can see in Fig. 5.4d, the GBPs grow bigger and are sporadically dispersed along the grain boundaries, accompanied by the existence of precipitation-free zones (PFZs). Consequently, the intergranular corrosion process in the RRA condition is further diminished, and only a few isolated corrosion pits are visible in localized areas along the grain boundaries without distinct continuous corrosion paths. In the case of the T73 conditioned sample, a thicker passivation film can be observed as shown in Fig. 5.3d, which exhibits only sporadic rupture and local dissolution with no signs of intergranular corrosion. In comparison to the RRA condition, GBPs significantly increase in coarseness, and the gaps between GBPs expand, resulting in a wider PFZ as shown in Fig. 5.4c. This observation indicates that the formation of intergranular corrosion paths in the electrochemical corrosion process becomes highly challenging, which also strongly suggests that the corrosion resistance in the T73 condition is the highest among all the heat treatments.

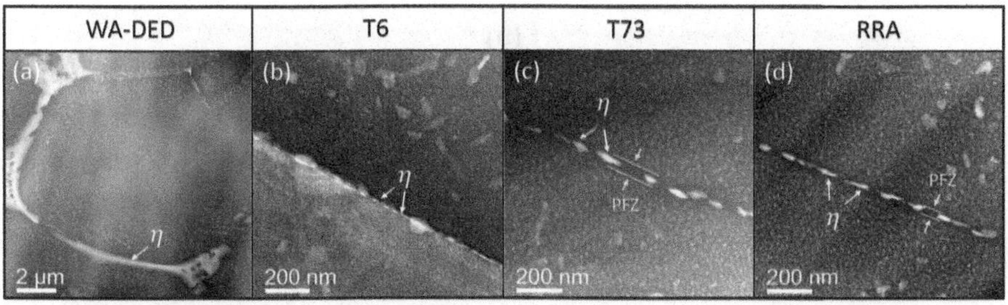

Fig. 5.4 STEM observations of GBPs under (**a**) WA-DED, (**b**) T6, (**c**) T73, (**d**) RRA conditions [9]

5.3.3 High-Cycle Fatigue Properties

In the previous discussion, aluminum alloys exhibit a significant tendency for porosity during the WA-DED process. Figure 5.5 shows the cross-sectional metallographic images of the 7B55-Sc alloy in the WA-DED condition. The biggest pore size in the WA-DED manufactured sample is around 690 μm. These porosity defects can seriously affect the fatigue performance of the material. Aircraft fuselage construction in the aerospace industry is largely accomplished using 7xxx series aluminum alloys. Hence, it is crucial to investigate the high-cycle fatigue performance of additively manufactured 7xxx aluminum alloys. Figure 5.6 presents the S–N curves of the 7B55-Sc alloy under both WA-DED and T6 heat-treated conditions. Regarding the WA-DED condition, the fatigue performance only reaches 100 MPa after 1×10^7 cycles. Apart from the presence of porosity defects, the persistent eutectic GBPs adversely affect the fatigue life of the material. This is primarily attributed to the recurring stress concentration in these areas during fatigue testing, leading to the initiation and rapid propagation of fatigue cracks along the brittle eutectic structures [17]. After the T6 conditioning, the fatigue performance of the 7B55-Sc alloy increases to 140 MPa. While the porosity defects still exist, those original coarse GBPs and intragranular second phases in the 7B55-Sc alloy are almost dissolved into the aluminum matrix, and subsequently reprecipitating as finer precipitates, resulting in an enhancement of fatigue performance to some extent.

Figure 5.7 shows the fatigue fracture surfaces of the 7B55-Sc alloys in the WA-DED and T6 heat-treated conditions. It is evident that fatigue initiation occurs at the locations of the porosity defects. This serves as clear evidence of the significant impact of porosity on the fatigue performance of the material, particularly as the high-cycle fatigue life often depends on the presence of material defects [18]. Therefore, addressing porosity defects is essential for additive manufacturing of aluminum alloys. For now, general methods to address porosity defects in additively manufactured aluminum alloys include interlayer deformation techniques, such as interlayer rolling and interlayer hammering, etc. [19–23]. These methods can help reduce some porosity defects in certain aluminum alloys, including 2xxx and 5xxx, to a certain extent due to sufficient deformation. However, the aforementioned

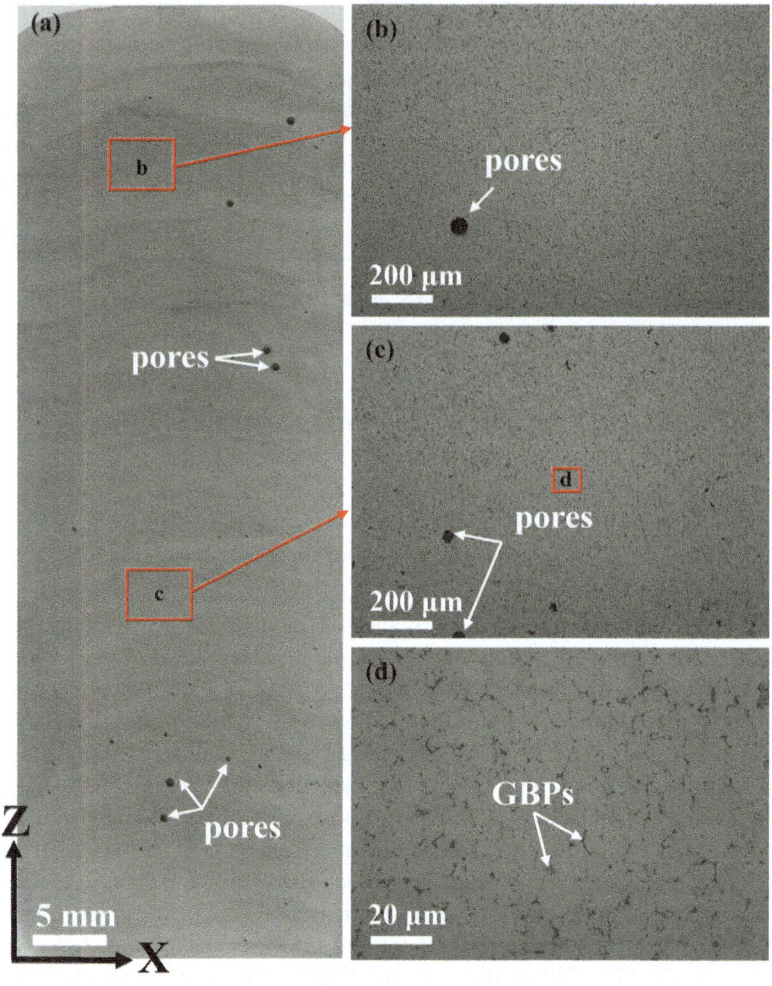

Fig. 5.5 Metallographic images of the WA-DED manufactured 7B55-Sc [13]

interlayer deformation methods are unable to address porosity defects in 7xxx aluminum alloys because the strain hardening in 7xxx aluminum alloys is too strong. In fact, insufficient deformation can even lead to the flattening of pores, resulting in even greater stress concentration at the pore tips, which further deteriorates the mechanical performance. Currently, an effective method for addressing porosity defects in additive manufacturing of 7xxx aluminum alloys is a novel hybrid process called the interlayer friction stir processing (FSP) technique. It is a cyclic process that enables interlayer FSP stirring after depositing a certain number of layers. This method introduces significant plastic deformation through the interlayer FSP process, thereby achieving both grain refinement and elimination of porosity defects [13], which can be treated as an effective solution for the future practical engineering applications of 7xxx aluminum alloys.

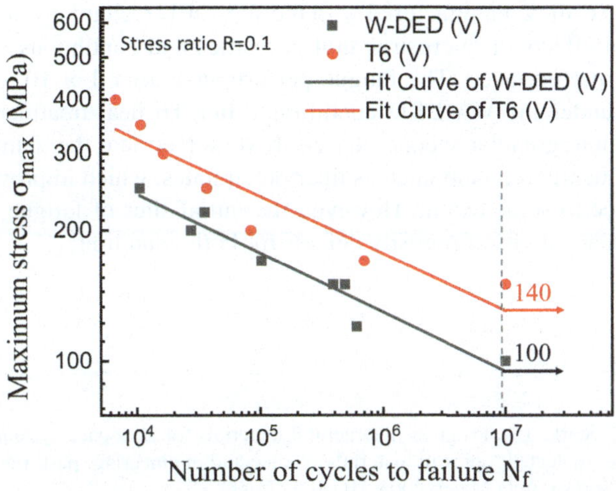

Fig. 5.6 S–N curves of the 7B55-Sc under WA-DED and T6 heat-treated conditions [13]

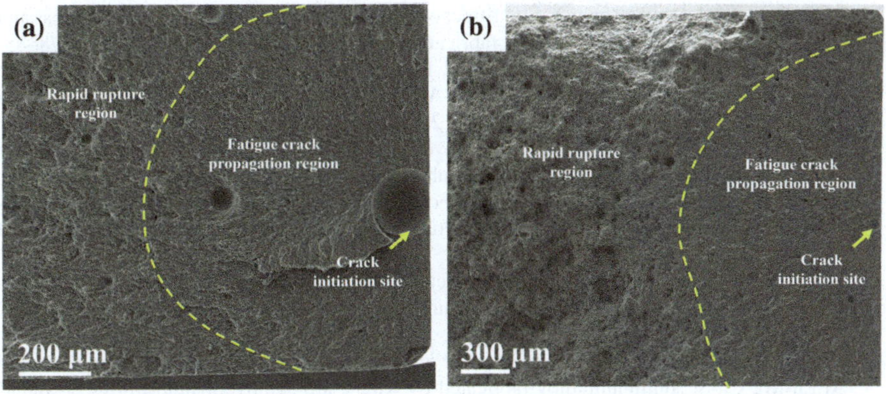

Fig. 5.7 Fatigue fracture surfaces of WA-DED and T6 heat-treated samples: (**a**) WA-DED under maximum stress of 120 MPa, (**b**) T6 under maximum stress of 170 MPa [13]

5.4 Conclusion

This chapter mainly investigates the corrosion resistance and fatigue performance of an aerospace material, specifically the 7xxx aluminum alloy manufactured by WA-DED technology. The key conclusions are as follows:

- The WA-DED process of the 7B55-Sc alloy exhibits the poorest corrosion resistance due to the presence of a large amount of continuous eutectic phases at grain boundaries, making it prone to the formation of intergranular corrosion paths. However, after undergoing the T73 overaging treatment, the GBPs become discontinuous, and the presence of PFZs prevents the extension of the corrosion path, resulting in significantly improved corrosion resistance.

- The 7B55-Sc alloy manufactured with the WA-DED technology suffers from the detrimental effects of microstructural porosities and continuous eutectic GBPs on fatigue performance. The fatigue performance after 1×10^7 cycles is only 100 MPa under the WA-DED condition. After T6 heat treatment, the coarse GBPs and intragranular second phases are dissolved into the aluminum matrix, and subsequently reprecipitated as finer precipitates, which improves the fatigue performance to some extent. However, the initial sites of fatigue fractures still occur at the locations of porosity defects for both conditions.

References

1. Williams JC, Starke EA. Progress in structural materials for aerospace systems11The Golden Jubilee issue—selected topics in materials science and engineering: past, present and future, edited by S Suresh. Acta Mater. 2003;51(19):5775–99.
2. Heinz A, Haszler A, Keidel C, Moldenhauer S, Benedictus R, Miller WS. Recent development in aluminum alloys for aerospace applications. Mater Sci Eng A Struct Mater Proper Microstruct Process. 2000;280(1):102–7.
3. Liu P, Hu J, Li H, Sun S, Zhang Y, bin. Effect of heat treatment on microstructure, hardness and corrosion resistance of 7075 Al alloys fabricated by SLM. J Manuf Process. 2020;60:578–85.
4. Klein T, Schnall M, Gomes B, Warczok P, Fleischhacker D, Morais PJ. Wire-arc additive manufacturing of a novel high-performance Al-Zn-Mg-Cu alloy: processing, characterization and feasibility demonstration. Addit Manuf. 2021;37:101663.
5. Klein T, Reiter L, Schnall M. Wire-arc additive manufacturing of Al-Zn5.5-Mg-Cu (ML7075): shifting paradigms of additive manufacture-ability. Mater Lett. 2022;313:131841.
6. Liu D, Wu D, Wang R, Shi J, Niu F, Ma G. Formation mechanism of Al-Zn-Mg-Cu alloy fabricated by laser-arc hybrid additive manufacturing: microstructure evaluation and mechanical properties. Addit Manuf. 2022;50:102554.
7. Fu R, Lu W, Guo Y, Lei H, Cui Y, Wang J, et al. Achieving high strength-ductility of Al-Zn-Mg-Cu alloys via hot-wire arc additive manufacturing enabled by strengthening precipitates. Addit Manuf. 2022;58:103042.
8. Guo X, Li H, Pan Z, Zhou S. Microstructure and mechanical properties of ultra-high strength Al-Zn-Mg-Cu-Sc aluminum alloy fabricated by wire + arc additive manufacturing. J Manuf Process. 2022;79:576–86.
9. Guo X, Xue P, Li H, Xu R, Ni D, Pan Z, et al. Microstructures and properties of wire-arc additively manufactured ultra-high strength aluminum alloy under different heat treatments. J Mater Res Technol. 2023;24:1663–78.
10. Chen S, Chen K, Peng G, Jia L, Dong P. Effect of heat treatment on strength, exfoliation corrosion and electrochemical behavior of 7085 aluminum alloy. Mater Design. 2012;35:93–8.
11. Li J, Peng Z, Li Z, Jia Z, Chen W, Zheng Z. Mechanical properties, corrosion behaviors and microstructures of 7075 aluminum alloy with various aging treatments. Trans Nonferrous Metals Soc China. 2008;18:755–62.
12. Guo X, Li H, Xue P, Pan Z, Xu R, Ni D, et al. Microstructure and mechanical properties of 600 MPa grade ultra-high strength aluminum alloy fabricated by wire-arc additive manufacturing. J Mater Sci Technol. 2023;149:56–66.
13. Guo X, Ni D, Li H, Xue P, Xu R, Pan Z, et al. Enhancing strength, ductility, and fatigue performance of Al-Zn-Mg-Cu-Sc-Zr alloy using a hybrid approach: wire-arc directed energy deposition and interlayer friction stir processing. J Mater Process Technol. 2023;322:118173.
14. Li S, Dong H, Shi L, Li P, Ye F. Corrosion behavior and mechanical properties of Al-Zn-Mg aluminum alloy weld. Corros Sci. 2017;123:243–55.

15. Wang C, Jiang F, Wang F. The characterization and corrosion resistance of cerium chemical conversion coatings for 304 stainless steels. Corros Sci. 2004;46:75–89.
16. Kim SJ, Jang SK. Effects of solution heat treatment on corrosion resistance of 5083F Al alloy. Trans Nonferrous Metals Soc China. 2009;19:887–91.
17. Carter ST, Rotella J, Agyei RF, Xiao X, Sangid MD. Measuring fatigue crack deflections via cracking of constituent particles in AA7050 via in situ x-ray synchrotron-based micro-tomography. Int J Fatigue. 2018;116:490–504.
18. Wang QY, Berard JY, Rathery S, Bathias C. High-cycle fatigue crack initiation and propagation behavior of high-strength spring steel wires. Fatigue Fract Eng Mater Struct. 1999;22:673–7.
19. Fang X, Zhang L, Chen G, Huang K, Xue F, Wang L, et al. Microstructure evolution of wire-arc additively manufactured 2319 aluminum alloy with interlayer hammering. Mater Sci Eng A Struct Mater Proper Microstruct Process. 2021;800:140168.
20. Gu J, Ding J, Williams SW, Gu H, Ma P, Zhai Y. The effect of inter-layer cold working and post-deposition heat treatment on porosity in additively manufactured aluminum alloys. J Mater Process Technol. 2016;230:26–34.
21. Gu J, Wang X, Bai J, Ding J, Williams S, Zhai Y, et al. Deformation microstructures and strengthening mechanisms for the wire + arc additively manufactured Al-Mg4.5Mn alloy with inter-layer rolling. Mater Sci Eng A Struct Mater Proper Microstruct Process. 2018;712:292–301.
22. Xie C, Wu S, Yu Y, Zhang H, Hu Y, Zhang M, et al. Defect-correlated fatigue resistance of additively manufactured Al-Mg4.5Mn alloy with in situ micro-rolling. J Mater Process Technol. 2021;291:117039.
23. Wei J, He C, Zhao Y, Qie M, Qin G, Zuo L. Evolution of microstructure and properties in 2219 aluminum alloy produced by wire arc additive manufacturing assisted by interlayer friction stir processing. Mater Sci Eng A Struct Mater Proper Microstruct Process. 2023;868:144794.

Chapter 6
Stress Corrosion Cracking in Aircraft Alloys

Mohammad Rauf Sheikhi

6.1 Introduction

Stress corrosion cracking (SCC) is a typical challenge in the aeronautical sector that poses severe risks to structural capability and safety of aircraft components. This failure mechanism comes into play when the combined effect of tensile stress and a corrosive environment leads to the formation and growth of material cracks. As components are installed in aircraft exposed to different environmental conditions and mechanical loads, it is essential to understand and prevent the occurrence of SCC to have a reliable and durable structure. This complexity can be attributed to the fact that SCC is not only a material property-dependent phenomenon but is simultaneously influenced by material properties, environment, and mechanical loads. Aircraft alloys, famous for their high strength-to-weight ratio and superior mechanical properties, are prone to SCC in some circumstances. Hence, it is critical to dissect the factors that regulate SCC in aircraft alloys to enable the formulation of effective prevention techniques and the design of reliable engineering solutions. Over the last few years, the aircraft industry has seen a shift of focus on sustainability and the cost of manufacturing, elevating the demand for solid and lightweight materials that are immune to rusting. However, these lightweight concepts use advanced alloys that probably have improved susceptibility to SCC under specific

M. R. Sheikhi (✉)
The State Key Laboratory of Heavy-duty and Express High-power Electric Locomotive, Central South University, Changsha, China

Key Laboratory of Traffic Safety on Track of Ministry of Education, School of Traffic & Transportation Engineering, Central South University, Changsha, Hunan, China

National & Local Joint Engineering Research Center of Safety Technology for Rail Vehicle, Central South University, Changsha, Hunan, China
e-mail: mohammadraufsheikhi@csu.edu.cn

S. Gürgen (ed.), *Failure in Aircraft Materials*, Sustainable Aviation,
https://doi.org/10.1007/978-3-031-65850-1_6

operational conditions. As a result, there is a growing research interest in understanding the SCC behavior of these new-age metallic materials for aircraft construction, such as aluminum alloys, titanium alloys, and high-strength steels.

6.2 Mechanisms in Stress Corrosion Cracking

The process of development of SCC starts with crack initiation. However, in some instances, there could be crack-like defects in the component itself, which behave as cracks and do not require a crack initiation step. Some initiation points are thus manufactural imperfections such as porosities, dents, laps, and grooves, while others are metallurgical imperfections like inclusions, phase changes, etc. In aircraft, fatigue is well implicated in SCC as cyclic loads create spaces for cracks. Depending on the environment, pitting corrosion, where initiation typically occurs at metallurgical discontinuities such as grain boundaries and inclusions, is a common characteristic. Crevice corrosion, dealloying, and intergranular corrosion initiate SCC in the components. Slip bands can also become sites of crack initiation owing to the localized stress accumulation, making crack initiation in areas of segregated solutions and dislocations easy [1, 2].

After crack initiation, the next step that happens in SCC is the crack growth phase. Often, this phase is not apparent at the macroscopic level. While there are no obvious signs of corrosion, even at the cross-section levels, it is also possible to identify some manifestations of local plastic deformation at the crack areas. It is important to note that corrosion products are often located within the crack and on its surface. While they can easily be deformed under normal conditions, SCC can lead to brittle cracks. Intergranular and cleavage-like features that exhibit brittle behavior are visible in the case of SCC. While overload causes highly absorbent fracture surfaces, SCC causes relatively smooth intergranular or several stepped cleavage fracture surfaces [2].

Environmental factors play a massive role in the operations of SCC. Some factors that affect the resistance of alloys to SCC include temperature, oxygen, chemical content, electrical potential, humidity, and pH level [3]. For example, SCC enhances highly above 60 °C in austenitic stainless steels because of higher rates of crack formation. Furthermore, chloride ions greater than this concentration are capable of causing SCC in these steels. On the other hand, hydrogen-sensitive alloys suffer from SCC mechanisms that have a negative electrical potential [2].

At that level, four significant mechanisms cause the extension of cracks in materials: These processes include atom removal at the crack tip into solution, the motion of atoms at the crack tip by shearing, the separation of atoms at the crack tip by tensile loads, and the surface diffusion of the particles from the crack tip. While the last three do not necessitate specific environments for occurrence, injurious environments can advance SCC in alloys. Concerning these atomic-level processes, several mechanisms have been suggested for the cause of SCC in materials [2].

The dissolution-based mechanism is one of the proposed mechanisms that describes the behavior of the process occurring at the atomic level at the crack tip. This mechanism claims atoms are displaced along the crack front toward the applied stress. Anodic residues, part of the environment, help this atomic dissolution point. As per this approach, dissolution, the most aggressive process, is expected to occur mainly at the crack tips where stresses are high, and the medium is most aggressive. On the other hand, the regions near the crack tips undergo only a slight dissolution due to the lower stresses and the less corrosive environment [2, 4].

About the process of fracture in alloys, an adsorption mechanism suggests that adsorption affects the fracture properties of alloys through the change of the energy at the surface. This reduction in surface energy helps in SCC by causing the separation of inter-atomic forces, making crack initiation easier on exposure to moist conditions [5]. In this mechanism, dislocation nucleation is said to occur at the crack tip, and these dislocations run further to encourage void formation in the strain area. This process leads to fractures that occur along the grain or intergranular freedom [2, 6].

Hydrogen-based mechanism is one of the recognized models for many materials, such as magnesium alloys, aluminum alloys, titanium alloys, and high-strength steel. Hydrogen, due to its atomic size, gets between dislocations and accumulates at crack tips in the case of alloys. Hydrogen production occurs in these materials through electrochemical processes or chemical dissolving water molecules. This hydrogen accumulation makes the material brittle, leading to SCC in the alloys. In exceptional cases, some alloys can also be hydrogen-based SCC in moist air. The fracture features of both the SCC and hydrogen embrittlement are somewhat comparable, apart from the existence of corrosive products on the fracture surface [2].

The vacancy mechanism is the other model considered to rationalize the existence of SCC in alloys. In this mechanism, dissolutions or dealloying processes produce vacancies at crack tips. These vacancies then spread in the deformation region, enhancing the development of cracks in the material. This leads to a discontinuous crack propagation described by recurrent corrosion cycles [7]. Additionally, it is suggested that dislocation emission occurs due to the accumulation of vacancies in the initial few atomic layers at the crack tips [2].

The surface mobility mechanism explains the function of single atoms or molecules at the surface of a material through elastic stress applied in diffusion to start or extend a crack. Surface diffusion results in the formation of vacancies, which substantiate crack advancement at the stressed crack tips. The rate of surface diffusion is normally higher in materials than in a bulk diffusion process. Furthermore, a wide range of SCC failures, which reveal intergranular and transgranular fractography, is associated with the surface mobility mechanism [2].

The film-induced cleavage mechanism is characterized by the forming of thin, brittle films in the environment created by crack tips. They break like glass under load, enabling cracks to penetrate the substrate. Even where the crack is arrested and blunted at some distance, the cycle is repeated many times. Every cycle creates recognizable points on the fracture surface representing crack arrest sites. This is so because the film layer has to be well bonded onto the substrate for the crack to progress within the material. As the film layers accumulate in thickness and have brittle characteristics, crack propagation rates remain incredibly high and do not reduce in the substrate material. Slip bands also act as barriers to arrest the crack and, as a result, result in very high dislocation activities in the material [2].

Besides the above-mentioned mechanisms, some other models have been proposed to explain the phenomenon of SCC in materials. Previous works have acknowledged various interactions between dissolution, oxidation, and mechanical failure processes [8–10]. These models offer additional knowledge about the nature of SCC and help to develop more comprehensive views on the factors influencing it.

6.3 Stress Corrosion Cracking Characteristics in Prominent Aircraft Alloys

6.3.1 Magnesium Alloys

Magnesium alloy is uniquely positioned in aircraft applications due to its low density and high strength-to-density ratio. These alloys include mainly magnesium and other modifying elements, including aluminum, zinc, and manganese, and they are pretty promising for replacing more conventional materials used in aerospace engineering, including aluminum alloys. This means that they are thin or have lesser thickness, which makes them have lesser structural weight, thus enhancing the aircraft's efficiency and performance as far as fuel is concerned. Magnesium alloys have another advantage: they have prodigious amounts of specific strength and stiffness, which are considerable as they are essential for a plane demanding to withstand the extremely harsh conditions it encounters during its flight. However, issues such as vulnerability to corrosion are some of the primary causes that warrant considering suitable alloy designs and protective measures when using these products in airplanes.

Regarding corrosion in magnesium alloys, there are two main groups: Zinc-containing and zinc-free alloys [11]. Among all the elements, zinc stands out as being crucial in magnesium alloys, whereby it has chemistry with other elements, which in turn enables purification of the melt containing different impurities. As a result, ordinary zinc-containing magnesium alloys are of high purity and have low corrosion rates, making them suitable for high-end uses such as in the aerospace industry, even though they are costly compared to standard grade. On the other hand, the usual magnesium alloy is mainly made of aluminum and contains no zinc. In general, one has to determine the amount of impurities in these alloys to know their corrosion resistance. Compared with the magnesium alloys containing zinc, zinc-free magnesium alloys have much higher purity and possess better corrosion-resisting ability Corrosion resistance [2].

Several metallurgical solutions were proposed to minimize corrosion of Mg alloys in salty water, for instance, a 3% NaCl solution: suitable α-Mg matrix, reasonable second phase quantity and distribution, and satisfactory chemical composition. In 3% NaCl solutions, the influence of alloy addition on the corrosion protection of solid solution type Mg-alloy is less efficient than that of pure Mg [12, 13]. All second phases enhance the micro-galvanic corrosion mechanism in the α-Mg matrix to varying extents. Iron, cobalt, nickel, and copper are the primary elements that galvanically corrode with the magnesium matrix at the micro level. Therefore, the presence of these elements in the above-predetermined quantities qualifies them as contaminants [14, 15]. Alloys of magnesium-containing elements usually exhibit lower corrosion resistance as compared with pure metal. However, the second phase, also known as the high corrosion resistance phase, can minimize the corrosion rate in the magnesium matrix. These phases are particularly beneficial if they form continuous morphologies that function as a protective barrier to corrosive species [2, 15].

Pure magnesium has a higher resistance to corrosion than magnesium alloy, but it is also affected by SCC. Earlier works [16, 17] have analyzed the SCC behavior of pure magnesium in different aggressive environments, including NaCl and chloride-chromate. These investigations established that pure magnesium displays transgranular fractures in fine-grain materials. However, as grain size increases, a mixed mode is reported, where both transgranular and intergranular features are seen. They also mentioned that when atomic hydrogen is in the solid solution form, it tends to decrease the surface energy, leading to the formation of cleavage-like fractures in the material [2]. Merson et al. [18] claimed that the adsorption mechanism will lead to the transgranular fracture morphology in pure magnesium, as depicted in Fig. 6.1.

Aluminum-containing magnesium alloys are particularly vulnerable to SCC, even in the presence of air. Substances, including distilled water and chloride, enhance the rate of occurrence of SCC in these alloys [2]. For instance, Busk [19], observed that the magnitude of SCC increases with the aluminum content in magnesium alloys, with an elevation from 1% to 8%. It is also advisable that the SCC threshold load in distilled water is about fifty percent of the yield strength for magnesium-aluminum alloys such as AZ91, AM60, and AS41 [2]. Chen et al. [20] experienced that the AZ91 alloy displays transgranular fracture behavior in an $MgCl_2$ solution where the threshold stress is reported to be around one-third of the yield stress. Padekar et al. [21] showed the SCC behavior of sand-cast magnesium alloy AZ91E in distilled water. The fractographies shown in Fig. 6.2 reveal that the alloy under investigation undergoes transgranular failure with parallel cleavage planes. Furthermore, there are secondary cracks present on the fractographs, which are marked by arrows in Fig. 6.2a. The fact that the fracture mode is transgranular implies that cracks propagated in a discontinuous manner in the alloy, a fact that is consistent with the hydrogen-based SCC mechanism.

Adding rare earth elements to magnesium-zinc alloys yields SCC resistance over magnesium-aluminum alloys. This improvement in corrosion resistance is due to

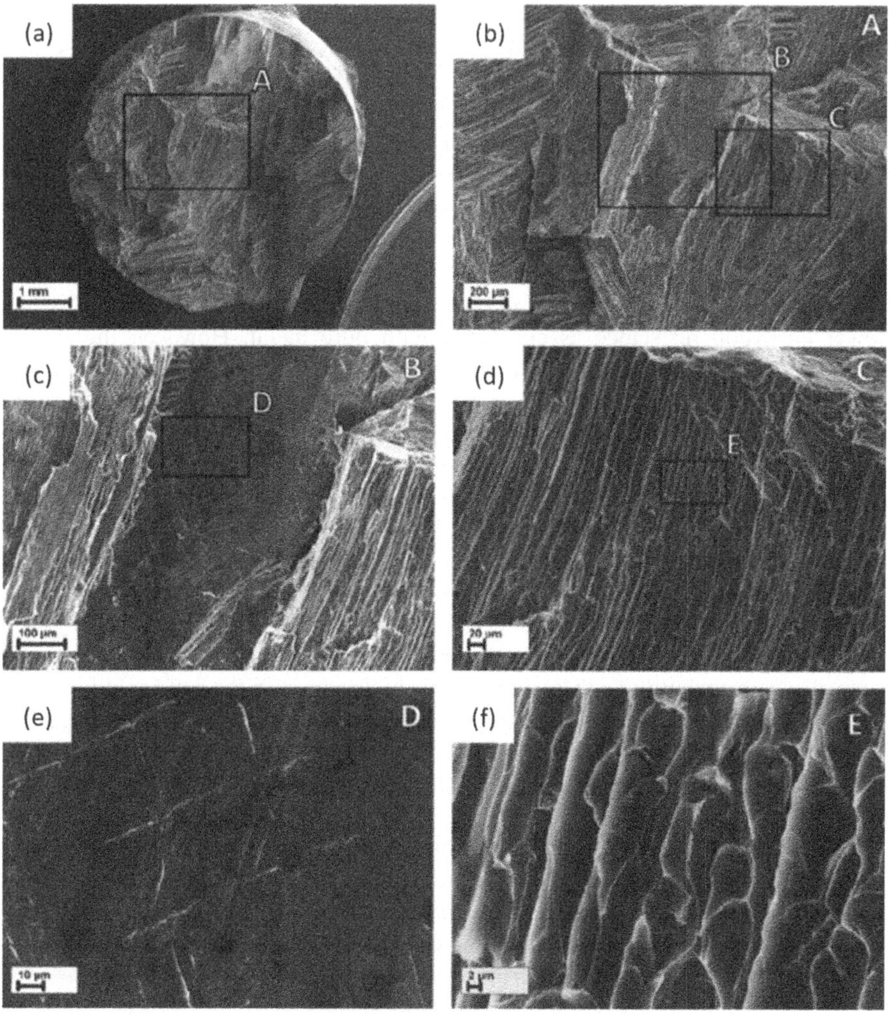

Fig. 6.1 Fracture surface of as-cast pure magnesium after SCC [18]

the formation of rare-earth-based phases at the grain boundaries, thus acting as a barrier to corrosion. Nonetheless, alloying magnesium-zinc with aluminum raises the corrosion rate, degrading SCC performance in both air and chloride-containing solutions. The SCC resistance is more excellent in magnesium manganese alloys when compared to air and chloride-based solutions. However, these alloys are unsuitable for application in distilled water or media containing sulfate ions in terms of SCC resistance [2].

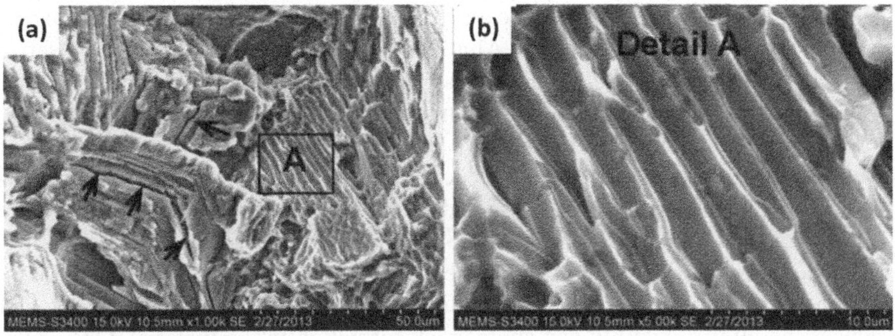

Fig. 6.2 Fracture surface of AZ91E after SCC in distilled water [21]

6.3.2 Aluminum Alloys

Aluminum alloys are effectively used in aircraft parts and equipment because of their low density, high strength, and high resistance to corrosion. These alloying agents include aluminum together with copper, magnesium, and zinc, and they help enhance the strength without compromising the weight of the aircraft, thus contributing to the decrease in mass of the aircraft while providing the much-needed structural and performance characteristics. Aluminum alloys like the 2000, 6000, and 7000 series used in the aerospace industry are classified by properties like durability and fatigue endurance. This difference makes aluminum alloys have better corrosion resistance than steel, making their use suitable for harsh weather conditions for aircraft, hence reducing maintenance costs.

Airplane parts are primarily manufactured from aluminum alloys; however, these parts are occasionally subjected to tensile loads and corrosive settings, and thus, SCC in aluminum alloys is an essential area of research. It was found that the SCC behavior of aluminum alloys depends on the contents of alloying elements, the state and distribution of precipitates in grain boundaries, and the density of dislocations. The following strategies have been employed to improve the SCC resistance in aluminum alloys: One such measure is heat treatment, which changes the structure of metals and removes stresses that contribute to SCC. Over-aging therapies are also employed to increase SCC resistance in high-strength aluminum alloys while simultaneously causing the loss in strength by making more considerable residues, which have lesser chances of getting stressed in service. Another way of enhancing SCC behavior in aluminum alloys is through retrogression and re-aging (RRA) treatment. It involves heating the alloy to an exceptionally high temperature for a short time (retrogression) before aging at a lower temperature. RRA treatment is especially suitable for the thin cross-sectional area and is familiar with sheet materials [2].

Various reports reveal that SCC occurs based on anodic dissolution and hydrogen-based mechanisms in high-strength Al alloys [2, 22, 23]. However, the 2000 series alloys are more susceptible to anodic dissolution-assisted SCC. This type of SCC strikes when the anodic oxidation of the material at the crack tip is enhanced by the corrosive environment, leading to crack extension. Nevertheless, the presence of copper in the 2000 series of alloys can encourage this form of corrosion as the localized anodic sites are put under stress and subjected to dissolution. In contrast, the 7000 series of aluminum alloys could be susceptible to hydrogen-induced cracking. This mechanism works through the intake of hydrogen in the metal and can build up in front of the crack and the metal lattice. Therefore, the addition of hydrogen reduces the alloy's ductility and spurs crack formation in the material. One of the features of the 7000 series Al alloys, which has been attributed to their high strength arising from additives such as zinc, is that they are very vulnerable to this form of SCC. Hydrogen-induced cracking in these alloys is preferential at the grain boundaries or by transgranular cleavage, thus revealing brittle failure modes.

The failure through anodic dissolution signifies lower resistance to crack formation and development, and in the case of aluminum alloys, this type of failure leads to intergranular fracture [22]. In this failure mode, the grain boundaries dissolve preferentially relative to other parts of the microstructure, resulting in a higher dissolution rate for the points labeled 'A.' The phenomenon of selective dissolution is attributed to the fact that the grain boundaries of the 2000 series aluminum alloys, which are copper lean or copper depleted, possess different electrochemical propensity than the copper-rich grain interior. This potential difference encourages intergranular corrosion throughout the microstructure of the plant absorbs currents. Subsequently, under tensile loading, the formation of such corroded pathways leads to cracking and continues further along these corroded paths, and in the alloys, leads to intergranular fracture. This mechanism is quite noteworthy, especially regarding the 2000 aluminum alloy family, because they affect the grain boundary electrochemistry compared to the grain [2].

As intimated in [24], initial research on Al alloys proved that intermetallic precipitates behave as preferred anode sites for SCC formation. It is also possible to weaken the alloy by dissolving these precipitates and opening the crack initiation point in the microstructure. This process is well illustrated in Fig. 6.3, which

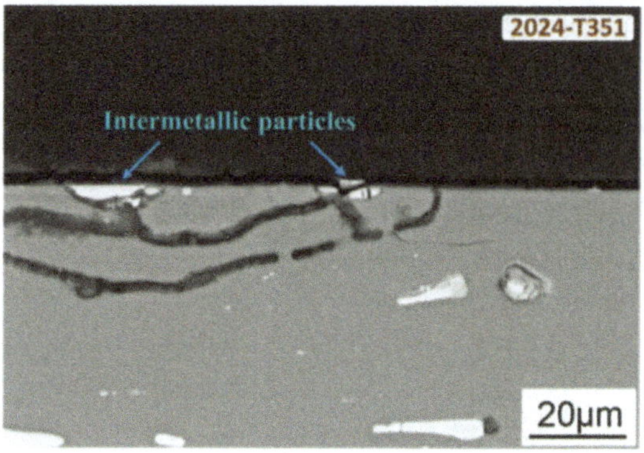

Fig. 6.3 Cross-sectional morphology of AA2024-T351 after exposing corrosive medium [25]

displays the cross-sectional microstructure of the AA2024-T351 aluminum alloy after immersion in a corrosive environment.

Another SCC mechanism in aluminum material is hydrogen-related cracking, where hydrogen ions are inserted into the aluminum alloys, especially the 7000 series. Dislocations, grain boundaries, or vacancies are some types of microstructural features that are assumed to act as traps that capture hydrogen atoms that are able to diffuse into the alloys. Loading hydrogen, particularly along the grain boundaries, further enhances the degree of embrittlement of aluminum alloys [2]. This usually leads to transgranular failure as the cohesion between the grains decreases along the grain boundaries. Intergranular failures are also possible, but they are limited to a few percentages only. However, another problem for the 7000 series alloys is magnesium segregation. There is also hydrogen embrittlement. Like hydrogen atoms, magnesium tends to reside at grain boundaries and leads to transgranular fractures. Figure 6.4 demonstrates how magnesium and hydrogen charging impact the grain boundary fracture stress. The chart illustrates that the fracture strength is reduced when the concentrations of magnesium and hydrogen are high at the grain boundaries. In these cases, magnesium segregation has a lesser impact on the SCC mechanism than hydrogen because magnesium's fracture strength is higher at concentrations similar to hydrogen. Also, magnesium has favorable compatibility with hydrogen; therefore, charging magnesium at the grain boundaries will cause an increase in the hydrogen solution. Consequently, magnesium segregation triggers additional hydrogen attack at the grain boundaries and thus enhances the SCC vulnerability in the 7000 series aluminum alloys [26].

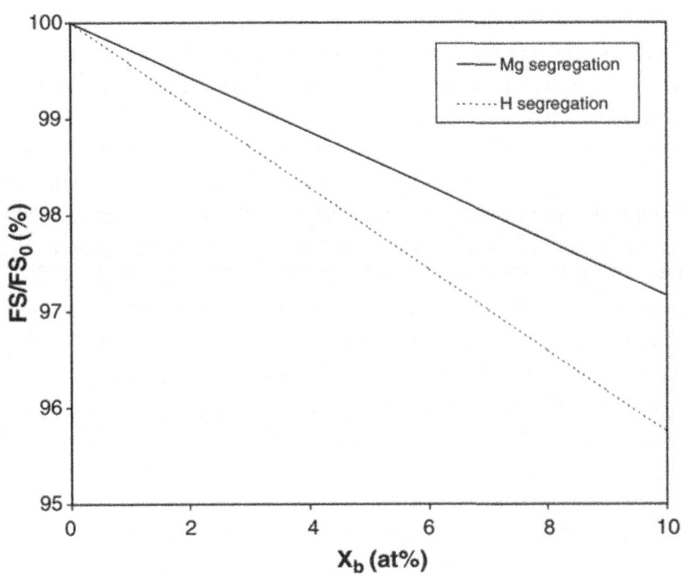

Fig. 6.4 Effects of magnesium and hydrogen charging on grain boundary fracture stress [26]

6.3.3 Titanium Alloys

Titanium alloy is essential in aircraft applications because it possesses high strength, low density, and corrosion resistance. These alloys are mainly based on titanium integrated with specific quantities of aluminum and vanadium. They are used to attain a satisfactory strength-to-weight ratio of enormous importance to increment aircraft efficiency. Titanium alloys also demonstrate a high degree of fatigue and fracture resistance to material, which is essential for the structural parts, such as landing gears and frames and engine parts that receive high stress and variable temperature exposure. These components also have excellent corrosion resistance, which is especially important in harsh environments, further increasing their service life and reducing requirements for maintenance.

Titanium alloys form oxide layers at their outer surfaces, capable of withstanding corrosive influences. Depending on the oxidation state, these protective oxide films typically contain TiO_2, Ti_2O_3, or TiO. The thickness of the film layer, usually about 10 nm, will be adequate to shield the alloy from conditions such as attack by hydrogen fluoride. Thus, an irritated or acid environment does not significantly influence the corrosion properties of titanium alloys when used up to the temperature of 70 °C. Therefore, in some cases, for example, in the presence of moisture, an alloy layer of passive film TiH_2 may be formed. This layer is also effective in preventing corrosion; however, there is a decrease in the material's flexibility. Also, the alloy may experience hydrogen charging, leading to hydrogen embrittlement [1, 2].

Though titanium alloys demonstrate good resistance to fracture and corrosion, cases of SCC have been noticed in such materials. The wear off of the protective film is the initial process in establishing SCC in these alloys. Because of the self-healing properties of the protective layers, the breakdown of the layers initiates cracks and influences the electrochemical kinetics [2]. The literature review [27, 28] reported that protective films can cause up to 60 MPa tensile stress, which significantly contributes to the SCC tendencies in the alloys.

For example, it was reported that secondary cracks could form in Ti-6Al-4V, widely used in aerospace industries, due to dissolution along the slip planes under SCC conditions. Twin Advances first disrupt the outer cover layers and stimulate the slip-line atoms in the dislocation movements. This facilitates preferential dissolution along the slip lines. They emanate from the crack tip and cause thinning locally, which in turn promotes the crack propagation in the alloy. When the applied tensile stresses are low, the stresses caused by the protective film may lead to crack growth. Further, hydride formations may be a significant cracking source due to the hydrogen effect, particularly in commercially pure titanium [2].

The aging of the titanium alloy material considerably influences the SCC sensitivity of single- and dual-phase titanium alloy. If the alloy contains a smaller amount of the α-phase, it demonstrates substantial resistance towards SCC in the NaCl environment. However, the β-phase alloys with greater aging durations falter in a transgranular fashion, then change to intergranular mode as they progress through the microstructure. Interstitial elements in titanium alloys also affect the tendency of materials to develop SCC. These concentrations make SCC more susceptible to NaCl solutions where low oxygen and high iron are realized. An increase in the oxygen concentration lowers the alloys' resistance to SCC in alloys containing low iron concentrations [1, 2].

Furthermore, texture contributes to the SCC susceptibility in titanium alloys since grain size/shape within the microstructure matters. For example, the SCC susceptibility of Ti-6Al-4V depends on the tendencies of the crystallographic planes that are sensitive to cleavage. As such, the microstructure constituents such as phase constitution, inter-crystalline element accumulation, and grain features decidedly influence the susceptibilities of titanium alloys to SCC [2].

Cao et al. [29] studied the SCC behavior of a near-α alloy, Ti-8Al-1Mo-1V, a member of dual-phase ($\alpha + \beta$) titanium alloys. The experimental studies were carried out on NaCl solutions. The findings of this investigation reveal that dislocation emission due to absorption and hydrogen-influenced localized plasticity also play a role in the occurrence of SCC in the alloy. A local increase of the dislocation density in the formation of the second crack is seen relative to the pre-crack sites, which attests to the absorption-induced dislocation emission causing microplastic deformation around the crack tip. Also, hydrogen enters the corrosive atmosphere, leading to hydrogen embrittlement around the crack tip, which is the cause of crack propagation through hydrogen in the alloy. Therefore, a combined SCC mechanism for the alloy Ti-8Al-1Mo-1V in NaCl solution can be recommended. Figure 6.5 provides a schematic of the alloy, revealing the areas where SCC facets are propagated in the material and the arrow showing the position of the dimpled area on the fracture surface.

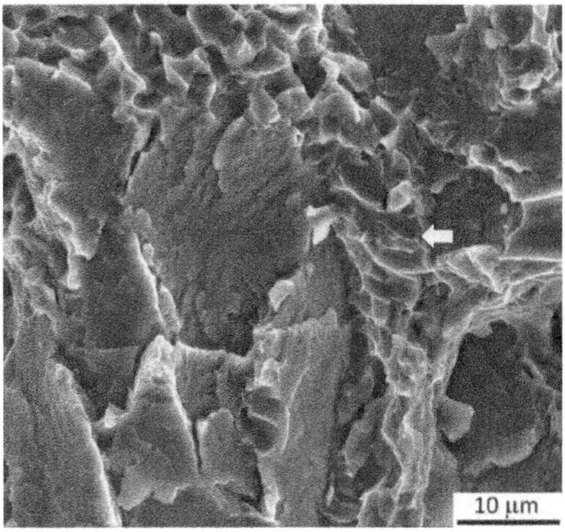

Fig. 6.5 Fracture surface of Ti-8Al-1Mo-1V in NaCl solution [29]

6.3.4 Stainless Steels

Stainless steels are essential in aircraft applications mainly because they have high corrosion, mechanical properties, and durability. These ferrous-based alloys with iron, chromium, and small additions of nickel and molybdenum contain a passive chromium oxide layer, giving high resistance to oxidation and corrosion, which is necessary for components used in aircraft construction operating in aggressive environments. These applications include fasteners, landing gear parts, exhaust systems, and structural parts, where strength and resistance to environmental degradation are essential. In addition to those fundamental characteristics of stainless steel, other mechanical properties like tensile strength and toughness can be improved using processes like heat treatment and cold working to produce parts with sufficient flightworthiness. Given this, it is impossible to overstate the importance of stainless steel as an essential material that will provide the necessary safety, durability, and efficiency of aeroframe elements for new-generation aircraft.

The increased resistance in corrosion for stainless steel is because of the creation of a chromium oxide film layer on the surface of the alloy. It forms a passive layer, which acts as a barrier to aggressive species in different environments and has a built-in capability of healing itself when it has undergone some deformation like scratching. Even though the SCC offers excellent protection against corrosion, the phenomenon can develop in stainless steel under conditions of tensile stress and exposure to a corrosive environment. However, as mentioned earlier, when the protective layer operates at SCC conditions, local breakdowns can occur due to local weakening mechanisms such as heterogeneities in the microstructure. This often results in pitting-type corrosion. This is true and results in pitting-type corrosion. However, SCC is not only caused by external tensile loads; residual stresses produced from welding or heat treatments can also lead to SCC in the alloys. Cracking attributable to SCC in stainless steel can be intergranular, transgranular, or intergranular-transgranular [2]. CSC in alloys is highly dependent on the corrosive environments that facilitate the formation of the SCC. This is true for most austenitic stainless steels where halides like NaCl constitute the most popular corrosive agents. It may be explained by the fact that chloride ions usually cause transgranular cracking with branches on the fracture surface. These two elements are fluorides, can readily dissolve into alloys in what may be referred to as ions, and create metal-fluoride bonds within the structure of the metallic material. In fluoride-based attacks, stainless steels are attacked uniformly except for pitting, which is not a characteristic of fluoride-based attacks. The Floridians are more active than the chlorine and, therefore, cause enhanced dissolution in the stainless steel.

An increase in temperature reduces the percentage of SCC in stainless steel. It is important to note that, at high temperatures, dissolution rates also rise, leading to faster rates of SCC in these alloys. Usually, SCC is not an apparent failure mode in stainless steel, predominantly at temperatures below 60 °C. Yet, as temperature increases, there is a buildup in the likelihood of SCC dominating and eventually becoming the leading cause of failure. SCC can be initiated without chloride ions, but pitting corrosion in chloride solutions causes SCC in stainless steels at even 50 °C [2, 30].

It is also important to note that boiling magnesium chloride is the most aggressive environment for stainless steel, primarily for the austenitic class. In this solution, SCC can emerge within a few hours without external loading, as otherwise, residual stresses in alloys contribute to the cracking. The solution has a boiling point of 155°C, and the aggressive effects reduce as the solution continues to boil since the solution becomes less acidic [31]. Also, when working with magnesium chloride solutions, the SCC tendency is much higher when more acid is added to decrease pH levels. Nevertheless, this pH reduction does not significantly impact the SCC behavior of stainless steels in calcium chloride solutions [2].

Ions like Fe^{3+} and Cr^{6+} are very effective in enhancing the formation of SCC in stainless steels and at room temperatures. Austenitic stainless steel is especially susceptible to SCC at ambient temperatures in chloride solutions containing ferric ions and fails rapidly [2]. The effect of oxidizing ions on the corrosion of stainless steels in nitric acid solution was studied by Liu et al. [32]. The authors identified that ions such as Cr^{6+}, V^{5+}, and Ce^{4+} affect anodic and cathodic processes and the properties of protective films on stainless steels. Figure 6.6 illustrates the influence of diverse oxidizing ions on the passive film thickness formed on the specimens.

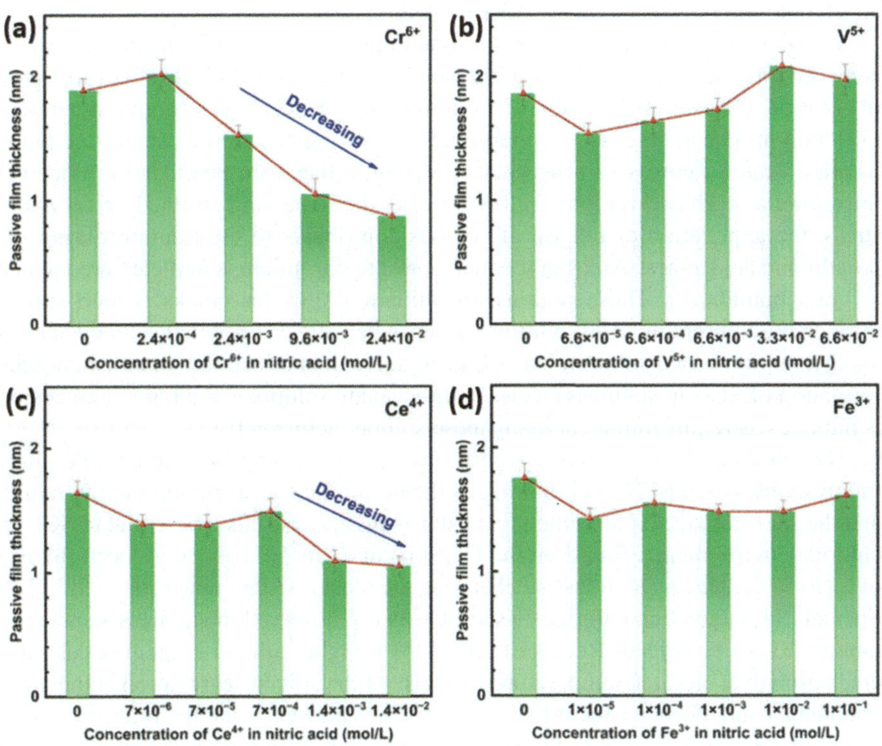

Fig. 6.6 Change in the passive film thicknesses with the oxidizing ions of (**a**) Cr^{6+}, (**b**) V^{5+}, (**c**) Ce^{4+}, and (**d**) Fe^{3+} in hot nitric acid solution [32]

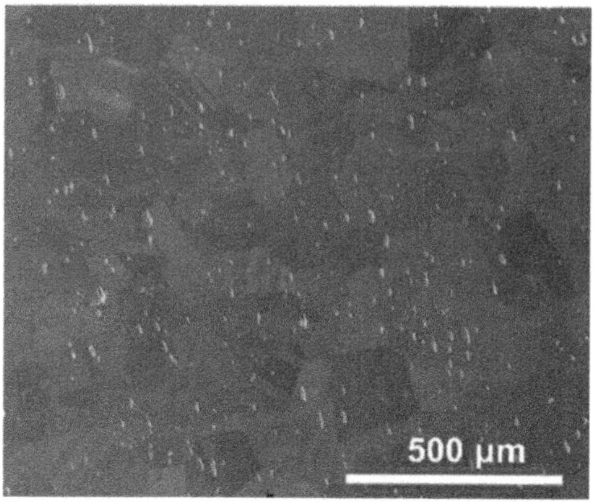

Fig. 6.7 Microstructural view of the specimen after the nitric acid exposure for 120 h [32]

That the number of film layers significantly decreases in conditions where the concentration of Cr^{6+} and Ce^{4+} in the nitric acid solutions is higher. These ions result in extensive dissolution of the passive film layers, leaving the material vulnerable to corrosion. Figure 6.7 illustrates the microstructure of a specimen after a nitric acid exposure test of 120 h that indicates the pitting corrosion and Localized corrosive attack randomly occurred on the microstructure.

Sulfide inclusions are established as potent promoters of SCC in stainless steels and even minimal concentrations of thiosulphate, such as 0. As little as 0.1 to 1 ppm of chloride, for example, can cause the initiation of SCC at even room temperature [33]. Polythionic acid results in intergranular cracking on stainless steels due to the increased rate of corrosive processes occurring at the grain boundaries. Since the intergranular corroded regions are bounded by the remaining strained layers under stress, the application or release of stresses contributes to the fracture during the polythionic acid attack. Areas in the matrix where chromium is depleted are located at grain boundaries. These regions are initiation sites for cracks, which spread through the intergranular dissolution regions when stress is applied. Thiocyanate is another sulfide-based solution but is less effective than thiosulfate in enhancing the formation of SCC in stainless steels. In thiocyanate solutions, the type of corrosion is pitting, where pits initiate at manganese sulfide inclusion [2].

The alloying elements play a significant role in forming SCC in straight chromium stainless steels. Nickel minimizes the austenitic transformation temperature, and the stainless steel containing nickel shows higher resistance potential to SCC in chloride environments. Based on the Copson curve, nickel has a high performance of up to 50 %, placing stainless steel among the most resistant materials to SCC [2]. Speidel [34] stated that chloride-based SCC is not observed in stainless steels containing 30–40 % Nickel. Recent studies [35] a higher silicon content accelerates crack growth at elevated temperatures because of the silicon segregation at the grain boundaries and the subsequent formation of silica phases inside the microstructure. Based on the data reported in this paper, the presence or absence of a molybdenum effect on SCC behavior depends on the context. In sodium chloride solutions, the

SCC resistance of stainless steel increases with the increase in the levels of molybdenum content. On the other hand, the effect of molybdenum on SCC resistance in magnesium chloride solutions is not so straightforward. At the same time, there is a tendency for SCC resistance to reduce as molybdenum content increases up to a certain level; beyond this level, SCC resistance increases with an increase in molybdenum concentration [2]. The influence of nitrogen on the resistance of SCC in stainless steels is generally favorable. Nitrogen-containing stainless steel exhibits a lower susceptibility to SCC as nitrogen promotes the formation of stable layers of protective films on the steel surface [2, 36, 37]

6.4 Conclusion

SCC is critical for aeronautical vehicles since it occurs in all main alloy systems utilized in their structure that must withstand various environmental and mechanical stresses. The consequences of SCC failures result in significant loss due to additional maintenance, which includes repairs, replacements, and frequent checks of suspicious components. This can even lead to the loss of an aircraft. This chapter briefly discussed SCC in aircraft alloys, emphasizing its significance, the processes by which it occurs, and the implications of the phenomenon. Due to some of the physical characteristics of SCC in materials used in aviation, this overview sought to improve the understanding of this phenomenon to extend the knowledge base in this field.

References

1. Jones RH, editor. Stress-corrosion cracking. Materials Park: ASM International; 1992. 448 p
2. Raja VS, Shoji T, editors. Stress corrosion cracking: theory and practice. Oxford/Philadelphia: Woodhead Pub; 2011. 792 p. (Woodhead Publishing in materials)
3. Shipilov SA, editor. Environment-induced cracking of materials. Amsterdam/New York: Elsevier; 2008. 2 p
4. Hall MM. Film rupture model for aqueous stress corrosion cracking under constant and variable stress intensity factor. Corros Sci. 2009;51(2):225–33.
5. Rehbinder PA, Shchukin ED. Surface phenomena in solids during deformation and fracture processes. Prog Surf Sci. 1972;3:97–188.
6. Liu HW. A unified model of environment-assisted cracking. Acta Mater. 2008;56(16):4339–48.
7. Milne I, Ritchie RO, Karihaloo BL, editors. Comprehensive structural integrity. 1st ed. Amsterdam/Boston: Elsevier/Pergamon; 2003. 10 p
8. Hou J, Peng QJ, Sakaguchi K, Takeda Y, Kuniya J, Shoji T. Effect of hydrogen in Inconel Alloy 600 on corrosion in high temperature oxygenated water. Corros Sci. 2010;52(3):1098–101.
9. Jones RH, Baer DR, Danielson MJ, Vetrano JS. Role of Mg in the stress corrosion cracking of an Al-Mg alloy. Metall Mater Trans A. 2001 Jul;32(7):1699–711.
10. Yu JG, Luo JL, Norton PR. Electrochemical investigation of the effects of hydrogen on the stability of the passive film on iron. Electrochim Acta. 2002;47(10):1527–36.

11. Winzer N, Atrens A, Song G, Ghali E, Dietzel W, Kainer KU, et al. A critical review of the stress corrosion cracking (SCC) of magnesium alloys. Adv Eng Mater. 2005;7(8):659–93.
12. Song G, Atrens A. Understanding magnesium corrosion—a framework for improved alloy performance. Adv Eng Mater. 2003;5(12):837–58.
13. Song G, Atrens A. Recent insights into the mechanism of magnesium corrosion and research suggestions. Adv Eng Mater. 2007;9(3):177–83.
14. Liu M, Uggowitzer PJ, Nagasekhar AV, Schmutz P, Easton M, Song GL, et al. Calculated phase diagrams and the corrosion of die-cast Mg–Al alloys. Corros Sci. 2009;51(3):602–19.
15. Zhao MC, Liu M, Song G, Atrens A. Influence of the β-phase morphology on the corrosion of the Mg alloy AZ91. Corros Sci. 2008;50(7):1939–53.
16. Meletis EI, Hochman RF. Crystallography of stress corrosion cracking in pure magnesium. Corrosion. 1984;40(1):39–45.
17. Winzer N, Atrens A, Dietzel W, Song G, Kainer KU. Comparison of the linearly increasing stress test and the constant extension rate test in the evaluation of transgranular stress corrosion cracking of magnesium. Mater Sci Eng A. 2008;472(1–2):97–106.
18. Merson E, Poluyanov V, Myagkikh P, Merson D, Vinogradov A. Fractographic features of technically pure magnesium, AZ31 and ZK60 alloys subjected to stress corrosion cracking. Mater Sci Eng A. 2020;772:138744.
19. Busk RS. Magnesium products design. New York: M. Dekker; 1987. 544 p. (Mechanical engineering)
20. Chen J, Ai M, Wang J, Han EH, Ke W. Stress corrosion cracking behaviors of AZ91 magnesium alloy in deicer solutions using constant load. Mater Sci Eng A. 2009;515(1–2):79–84.
21. Padekar BS, Raja VS, Raman RKS, Lyon P. Stress corrosion cracking behavior of magnesium alloys EV31A and AZ91E. Mater Sci Eng A. 2013;583:169–76.
22. Bobby-Kannan M, Raja VS, Raman R, Mukhopadhyay AK. Influence of multistep aging on the stress corrosion cracking behavior of aluminum alloy 7010. Corrosion. 2003;59(10):03100881.
23. Bobby Kannan M, Raja VS. Hydrogen embrittlement susceptibility of over aged 7010 Al-alloy. J Mater Sci. 2006;41(17):5495–9.
24. Bobby Kannan M, Raja VS. Role of coarse intermetallic particles on the environmentally assisted cracking behavior of peak aged and over aged Al–Zn–Mg–Cu–Zr alloy during slow strain rate testing. J Mater Sci. 2007;42(14):5458–64.
25. Wang L, Liang J, Li H, Cheng L, Cui Z. Quantitative study of the corrosion evolution and stress corrosion cracking of high strength aluminum alloys in solution and thin electrolyte layer containing Cl-. Corros Sci. 2021;178:109076.
26. Song RG, Dietzel W, Zhang BJ, Liu WJ, Tseng MK, Atrens A. Stress corrosion cracking and hydrogen embrittlement of an Al–Zn–Mg–Cu alloy. Acta Mater. 2004;52(16):4727–43.
27. Guo XZ, Gao KW, Chu WY, Qiao LJ. Correlation between passive film-induced stress and stress corrosion cracking of α-Ti in a methanol solution at various potentials. Mater Sci Eng A. 2003;346(1–2):1–7.
28. Lu H, Gao KW, Qiao LJ, Wang YB, Chu WY. Stress corrosion cracking caused by passive film-induced tensile stress. Corrosion. 2000;56(11):1112–8.
29. Cao S, Zhu S, Samuel Lim CV, Zhou X, Chen X, Hinton BRW, et al. The mechanism of aqueous stress-corrosion cracking of α + β titanium alloys. Corros Sci. 2017;125:29–39.
30. Sedriks AJ. Corrosion of stainless steels. 2nd ed. New York: Wiley; 1996. 437 p. (The corrosion monograph series)
31. Maier IA, Manfredi C, Galvele JR. The stress corrosion cracking of an austenitic stainless steel in HCl + NaCl solutions at room temperature. Corros Sci. 1985;25(1):15–34.
32. Liu Z, Zhang LM, Chen WQ, Ma AL, Zheng Y, Yan W, et al. Effect of oxidizing ions on the corrosion behavior of SiN stainless steel in high-temperature nitric acid solution. Electrochim Acta. 2023;442:141917.
33. Watanabe Y, Kondo T. Current and potential fluctuation characteristics in Intergranular stress corrosion cracking processes of stainless steels. Corrosion. 2000;56(12):1250–5.

34. Speidel MO. Stress corrosion cracking of stainless steels in NaCl solutions. Metall Trans A. 1981;12(5):779–89.
35. Andresen PL, Morra MM. Effects of Si on SCC of irradiated and Unirradiated stainless steel and nickel alloys. NACE Corrosion. 2005;1:05591.
36. Bali SC, Kain V, Raja VS. Effect of low-temperature sensitization on Intergranular stress corrosion cracking behavior of austenitic stainless steels in simulated boiling water reactor environment. Corrosion. 2009;65(11):726–40.
37. Kain V, Samantaray R, Acharya S, De PK, Raja VS. Influence of low-temperature sensitization on stress corrosion cracking of 304LN stainless steels. In: Environment-induced cracking of materials [Internet]. Elsevier; 2008 [cited 2024 May 27]. p. 163–172. Available from: https://linkinghub.elsevier.com/retrieve/pii/B9780080446356500558

Chapter 7
Creep Failure in Aeroengine Components

Salomé Sanchez and Zhijun Qiu

7.1 Introduction

Aeroengines are the main propulsion system of an aircraft. They generate thrust by converting fuel into mechanical energy to propel the aircraft forward. Jet engines, and more specifically gas turbine jet engines, are some of the most commonly used aeroengines as they have high efficiency and power-to-weight ratio, thanks to numerous technological advancements in the last decades. Gas turbine jet engines have five key stages:

1. Air intake: Air is drawn through the engine and its pressure increases before reaching the compressor stage.
2. Compressor: A series of compressor blades compress the air, increasing its pressure and temperature before heading to the combustion stage.
3. Combustion: Fuel is injected with the compressed air in the combustion chamber before being ignited. The mixture combusts, and high-pressure, high-temperature air heads to the turbine stage.
4. Turbine: The high-pressure, high-temperature air goes through a series of turbine blades, which extract power used to generate thrust.
5. Exhaust: The hot air leaves the exhaust at high velocity and generates thrust.

S. Sanchez (✉)
Department of Design Production and Management, Faculty of Engineering Technology, University of Twente, Enschede, Netherlands
e-mail: s.sanchez@utwente.nl

Z. Qiu
Australian Nuclear Science and Technology Organisation (ANSTO), Sydney, NSW, Australia

School of Mechanical, Materials, Mechatronic and Biomedical Engineering, University of Wollongong, Wollongong, NSW, Australia

S. Gürgen (ed.), *Failure in Aircraft Materials*, Sustainable Aviation,
https://doi.org/10.1007/978-3-031-65850-1_7

Numerous aeroengine components, such as the combustion chamber and the turbine blades, operate at elevated temperatures and in high-stress environments. This means that some of the main causes of failure include hot corrosion, fatigue, creep, creep-fatigue, and erosion failure [1]. In such environments, superalloys are often used as they maintain high mechanical integrity at elevated temperatures (close to their melting point) and resist mechanical degradation [2].

Ni-based superalloys are a family of modern aerospace engine materials [3, 4] and have been widely used in critical engineering applications and jet engine applications [5]. They are often used in the hottest parts of the aeroengine, particularly as turbine blades, disks, seals, rings, and other components. Miller [6] even reported that Ni-based superalloys account for 50 wt% of materials used in aeroengines. There are various types of Ni-based superalloys, such as Nimonic alloys—which were one of the first Ni-based superalloys developed in the 1940s [3]—and the Inconel, Rene, Hastelloy, Waspelloy, and Udimet series.

Ni-based superalloys have numerous phases. The γ' phase is the main strengthening phase [7], and some detrimental phases include Topologically Close-Packed (TCP) phases (e.g., σ). The size and morphology of these phases evolve with time, temperature, and stress. Specific morphologies can either be advantageous or detrimental to mechanical properties. This microstructure composition means that Ni-based superalloys boast high strength and toughness, corrosion and oxidation resistance, and creep resistance. Because of these advantageous characteristics, these superalloys are also difficult to machine [8, 9] and are therefore usually manufactured using casting or forging techniques [10]. The performance of many aeroengines is often limited by the high-temperature strength of these superalloys [2], of which creep is one of the leading causes of failure. Single-crystal superalloys are renowned for their superior creep properties, and their microstructure is characterized by a coherently embedded secondary strengthening phase (e.g., γ' phase), acting as an obstacle to dislocation movement, in a large volume fraction of a face-centered cubic γ matrix. Thus, dislocation networking can be formed in the γ/γ' interface [11]. However, with long-term exposure to high temperatures under service, these single-crystal superalloys are also prone to creep.

Creep describes a high-temperature (above 0.3 to 0.4 times the melting temperature in Kelvin) and time-dependent continuous plastic deformation process under a constant load. Creep occurs when exposed to heat and stresses below the conventional yield strength for a long-term period. Therefore, creep strain (ϵ) is a function of time (t), temperature (T), and stress (σ). The creep phenomenon consists of three stages, namely primary, secondary, and tertiary creep. It should be mentioned that under certain conditions, there is no secondary stage, and the tertiary stage develops immediately after the primary stage. In the initial stage, also called the primary stage, the creep strain rate is relatively high and decreases with time because of work hardening of the material. This is caused by the movement of dislocations, as well as the creation of subgrains, leading to an increase in dislocation density and a decrease in grain size. During the secondary stage, the creep rate decreases and reaches its minimum value, resulting in steady-state creep. At this stage, diffusion occurs—the softening recovery rate compensates for the strain hardening rate—leading to a stable creep rate without net structural changes. However, under continuous loading, the creep rate increases with time and enters the tertiary stage, which is also called the accelerated creep stage. During the tertiary stage, an increase in creep deformation and an accelerated creep rate result in necking or cross-section reduction before fracture.

Complicated metallurgical mechanisms, including the formation and propagation of microvoids or microcracks and grain boundary slip, usually occur in the creep rupture of Ni-based superalloys, followed by transgranular or intergranular fracture. These differences in fracture morphology are reflected in the density of pores, dimple sizes, and crack propagation traces. Creep fractures can be wedge-type (under higher stress, lower temperatures, and larger grain sizes) or r-type, with irregularities or cavities along the grain boundaries. The formation of cavities can lead to the formation of microcracks with a pile-up of dislocations and stress concentrations.

Creep deformation in metals is often related to several mechanisms, which involve microstructural evolution and occur at various temperatures and stresses. These mechanisms can be classified into two main types: dislocation creep and diffusion creep. The movement of dislocations is important and is highly influenced by the applied stress. Blum and Eisenlohr [12] state that the creep of crystalline materials is a plastic deformation by dislocation motion, regardless of the complementary mechanisms of plastic deformation, including the movement of grain boundaries and diffusion. Nabarro-Herring is a diffusion behavior, which is usually strongly dependent on temperature and typically occurs under high temperatures and moderate loads. Another diffusional-type mechanism is Coble creep, during which the elongation of the grains occurs along the loading orientation because of grain boundary sliding. Compared with Nabarro-Herring, the materials fabricated with fine grain size are more susceptible to Coble creep.

To determine the creep mechanism in the investigated material, the stress exponent, n, is determined by Eq. 7.1, where A is a dimensionless material constant, D is the diffusion coefficient, G is the shear modulus at the testing temperature T (absolute temperature), b is the Burgers vector, k is the Boltzmann constant, d is the grain size, σ is the applied stress in the loading direction, and p is the inverse grain size exponent. Note that fundamentally $n = 1$ for diffusion-controlled creep (Nabarro-Herring), $n = 2$ for grain boundary sliding [13], and $n = 4.5$ for dislocation-controlled creep (five-power law creep). Therefore, based on various material parameters and fitting, the creep mechanism can be calculated and determined.

$$\dot{\varepsilon}_{\mathrm{c,min}} = \frac{ADGb}{kT}\left(\frac{b}{d}\right)^{p}\left(\frac{\sigma}{G}\right)^{n} \tag{7.1}$$

The creep properties and behavior of superalloys have been extensively researched, as highlighted by a search through the SCOPUS database, which revealed that over 5000 documents (e.g., journal papers and conference proceedings) fit that description. Additionally, 72% of the search results were about Ni-based superalloys, compared to only 5% about Ti-based superalloys. Due to the sensitive nature of aerospace and superalloy topics, this chapter focuses on available studies on the creep failure of aeroengine components and extends into a more general overview of the creep behavior of Ni-based superalloys. First, typical creep failure mechanisms in Ni-based superalloy aeroengine components are examined, followed by an outline of the effects of stress and temperatures on creep failure. Following this, solutions to enhance creep performance are discussed.

7.2 Creep Failure Mechanisms

There are a limited number of studies about actual aeroengine components since companies are not willing to divulge their trade secrets, and the application of these components (e.g., military and space applications) is often subject to export control and intellectual property regulations. Of the available studies, the creep failure of turbine blades has been the subject of numerous studies [14–18]. Wang et al. [16] cut creep samples from different locations of directionally solidified cast turbine blades of DZ125 and found that the creep properties of the blade body were worse than those of the tenon tooth area, highlighting the anisotropy of turbine blades. Additionally, Ejaz et al. [15] investigated the creep failure of a Udimet 500 low-pressure turbine blade (without additional coatings) at temperatures of 760–800 °C. Failure was initiated by a triple point creep crack near the trailing edge of the blade before fatigue crack growth took over and caused the final fracture [15]. In this case, the initiation of the failure of the component was caused by creep. This is also the case for other studied parts, such as Udimet 500 high-pressure turbine blades, where failure was initiated by intergranular cracks caused by high-temperature exposition, followed by fatigue [14]. Creep and fatigue are highly intertwined in aeroengine component failure, and creep-fatigue constitutes a field of its own, which will not be touched upon in this chapter.

Ni-based single-crystal superalloys have also been researched extensively, as highlighted in a review by Xia et al. [19], and are known for their high creep performance. The creep properties of single-crystal Ni-based superalloys are mainly controlled by the microstructure and its evolution during creep, which results in various failure mechanisms.

There are different types of creep failure mechanisms, which depend on microstructural activity. Some of the main features affecting the creep mechanisms of polycrystalline and single-crystal Ni-based superalloys include dislocation activity and microstructure (phase and precipitate) evolution.

Creep mechanisms are determined by dislocation activity in the microstructure of single-crystal [19] and polycrystalline Ni-based superalloys, such as disks [20]. The majority of the creep lifetime of Ni-based superalloys is spent in the tertiary creep stage [21], which is caused by the accumulation of dislocations [22]. Epishin et al. [23] observed three different mechanisms during high-temperature, low-stress creep:

- Transverse glide of dislocations normal to the load direction during primary creep.
- Transverse glide-climb of dislocations during secondary creep (and contributing to the majority of the creep strain in steady-state creep).
- Climb of dislocations, which appeared during secondary creep.

It was also noted that γ' dislocations acted as a vacancy sink, facilitating the transverse glide-climb of dislocations for [001] single-crystal superalloys CMSX-4 and CMSX-10 during creep at high temperature and low stress (1100 °C with stresses between 105 MPa and 135 MPa) [23].

As well as dislocation activity, precipitates present in Ni-based superalloys, such as γ', carbides, and TCP phases, significantly affect creep performance. Particularly, γ' plays a very important role, especially in single-crystal superalloys [24]. In the case of single-crystal Ni-based superalloys, their higher creep strength stems not only from their single-crystal nature but also from their γ/γ' microstructure, which determines the dislocation motion and creep mechanism [19]. Increasing the volume fraction of γ' is often done to enhance strengthening effects and creep performance. In a study by Murakumo et al. [24] on single-crystal superalloy TMS-75 with various volume fractions of γ' precipitate, the longest creep life at 900 °C and 1100 °C was obtained by an alloy with a γ' content of around 65%. Similar results were observed in a study on Inconel 713C [25]. For creep samples taken from various locations of a DZ125 cast directionally solidified turbine blade, it was found that the creep properties of the body of the blade were worse than those of the tenon tooth area due to the higher volume fraction of γ' in the tenon tooth area, as shown in Fig. 7.1 [16]. However, this difference in precipitate did not affect the failure mode, as both the tenon tooth and body of the blade showed the same fracture mechanism—i.e., nucleation, accumulation, and connection of microvoids leading to large dimples on the fracture surface [16]. Due to the advantageous properties of γ'-strengthened Ni-based superalloys, these are typically used for disk applications [20].

Despite the benefits brought by the presence of γ' precipitates, the coarsening of γ' precipitates and the rafting process are some of the main microstructural mechanisms, which lead to creep failure. The coarsening of γ' is greatly affected by the lattice misfit between γ and γ'. An adequate lattice misfit can promote the formation of γ', but a too-large lattice misfit promotes the coarsening of the γ' phase, resulting in weaker mechanical integrity at high temperatures [19]. The effect of γ' coarsening was observed in turbine blade studies, as well as the effect of carbides [14, 17, 26].

The formation of continuous grain boundary carbides and the degeneration of MC carbides were one of the main reasons for structural degradation in turbine blades [14] and also the cause of the difference in creep performance in different locations of an Inconel 939 turbine vane after long-term exposure [26]. Additionally, a significant increase in the minimum creep rate was observed in low-pressure turbine rotor blades of wrought AP220BD and was attributed to the degradation of grain boundary carbide morphology, as shown in Fig. 7.2 [17]. The presence of primary and secondary carbides can also impede creep performance, as was shown

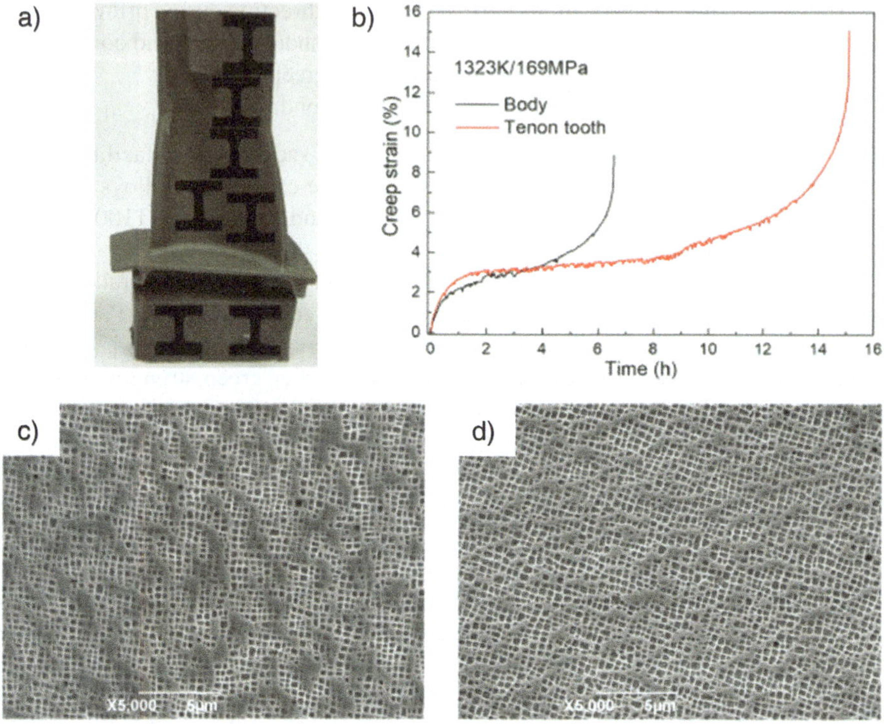

Fig. 7.1 (**a**) Overview of the creep specimens created from various locations of the blade, (**b**) creep curves for the body and tenon tooth, showing longer creep life for the tenon tooth area of the blade, (**c**) image of the γ' present in the body of the blade, and (**d**) image of the γ' present in the tenon tooth area of the blade [16]

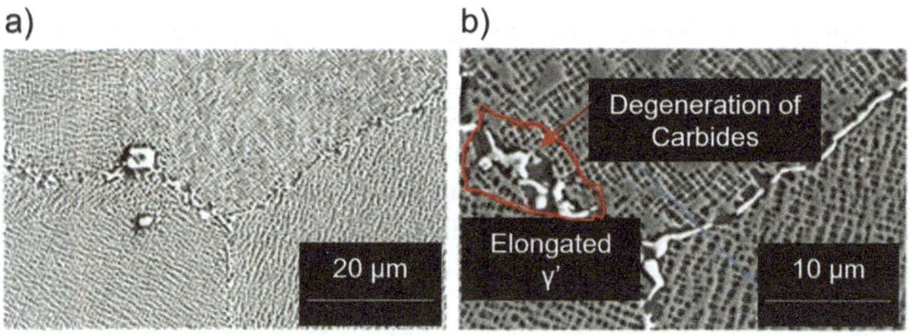

Fig. 7.2 (**a**) Grain boundary carbide in a 1600 h turbine blade and (**b**) morphology of a grain boundary carbide in a 1600 h turbine blade. Adapted from [17]

in a study on the Udimet 500 low-pressure turbine blade [15]. In wrought Inconel 718, the increase in the number of grain boundary carbides during creep served as microvoid initiation sites, which caused accelerated damage during tertiary creep and reduced creep life [27].

Apart from γ' precipitates and carbides, other phases and precipitates also impact creep performance.

This is the case for the TCP phase, which promotes the nucleation and propagation of microvoids, detrimental to creep properties. For example, the TCP, σ phase is a brittle intermetallic phase, which was also shown to cause structural degradation in turbine blades [14].

Overall, the various creep mechanisms are controlled by dislocation motion, which is significantly affected by temperature. At low temperatures, dislocation motion is restricted, and creep can only occur when high stresses are also present, whereas at high temperatures, creep can occur even at low stresses. This highlights the influence of temperature and stress on creep properties.

7.3 Influence of Temperature and Load on Creep

Temperature and stress both affect dislocation movements and, hence, different creep mechanisms. Higher stresses and temperatures result in a shorter creep life and a faster creep rate [28]. For the same temperature, higher stresses result in larger creep strain and creep rates, while for the same stress, a higher temperature results in a larger deformation [19]. Creep is usually categorized as "high temperature, low stress" or "low temperature, high stress." Although the definition of low/high temperature depends on the specific superalloy in relation to its melting point, low-temperature creep of Ni-based superalloys is usually around 750 °C, while mid-temperature creep is around 950 °C and high-temperature creep occurs when temperatures are over 1100 °C.

At low temperatures, for single-crystal superalloys, creep deformation is mainly controlled by dislocation activity with planar deformation and shearing of γ' by dislocations [20]. Smith et al. [29] also observed thermally activated shearing modes (e.g., microtwinning, stacking fault ribbons, and isolated superlattice extrinsic stacking faults) during low-temperature creep of both polycrystalline and single-crystal Ni-based superalloys. For high-temperature creep, the increased dislocation activity—namely dislocation climb and bypass [29]—results in an accumulation of creep strains [19].

Smith et al. [29] created a map of underlying creep mechanisms with respect to temperature and stress. The evolution of the microstructure during creep, such as the coarsening or rafting of γ' precipitates, causes differences in deformation mechanisms [19, 23, 29, 30]. Epishin et al. [23] observed that the γ' dislocations acted as a vacancy sink, which facilitated the transverse glide-climb of dislocations for [001] single-crystal superalloys CMSX-4 and CMSX-10 creep-tested at high temperature, low stress (1100 °C with stresses between 105 MPa and 135 MPa).

Numerous research papers have highlighted that temperature is more influential than stress in relation to creep performance [17, 31], mainly due to the influence of temperature on dislocation movement [19]. It was even shown that a rise in entry temperature in aeroengines resulted in faster creep rates and fatigue damage in high-pressure turbine blades [32]. However, single-crystal materials are anisotropic [33] and are greatly affected by the loading direction.

The effect of load direction seems more pronounced during low-temperature conditions, as was demonstrated by Wollgramm et al. [30] during miniature tests of single-crystal superalloy ERBO1 at precise [0 0 1], [1 1 0], and [1 1 1] loading directions. Indeed, a strong effect was observed during low-temperature creep when loading in different directions, whereas the effect was moderate at high temperatures [30]. Although the effect of loading direction may be less at high temperatures, it still influences the creep properties of the material. Nörtershäuser et al. [34] studied the high-temperature, low-stress creep of single-crystal superalloy LEK 94 in precise [001] and [110] loading directions. Loading in precise [110] showed significantly lower minimum creep rates, but even slight deviations resulted in a significant increase in the minimum creep rate. However, Heep et al. [35] compared different loading directions (precise [001] and 15° deviations from [001] towards [111] and [011]) during low-temperature (1023 K) high-stress (800 MPa) tensile creep of single-crystal ERBO-1 superalloy and determined that a 15° deviation could be tolerated, except towards the [111] direction.

Agudo Jácome et al. [36] explain that the variation of the minimum creep rate in the [001] direction is related to the activation of a greater number of crystallographic slip systems and the presence of dislocation networks at γ/γ' interfaces, which affect lattice misfit. During [110] loading, rafting and γ' cutting events influence the minimum creep rate [36, 37]. Rae et al. [38] also observed a high dependence between the creep strain rate and the applied stress during the primary creep stage of single-crystal CMSX-4 superalloy at 750–850 °C.

The morphology of creep fracture can also be influenced by the applied creep stress, and it was reported that the creep rupture mode changes from a mixed fracture to an intergranular fracture with increasing stress [39]. The mode of fracture may even vary in a single creep sample and show signs of mixed fracture. For example, Fig. 7.3 shows the various fracture morphologies observed by Zhao et al. [40] after creep testing of superalloy GH4169, including crack initiation (zone I), intergranular crack propagation (zone II) with signs of ductile failure (Fig. 7.3a, c), and intergranular brittle fracture (zone III) (Fig. 7.3c).

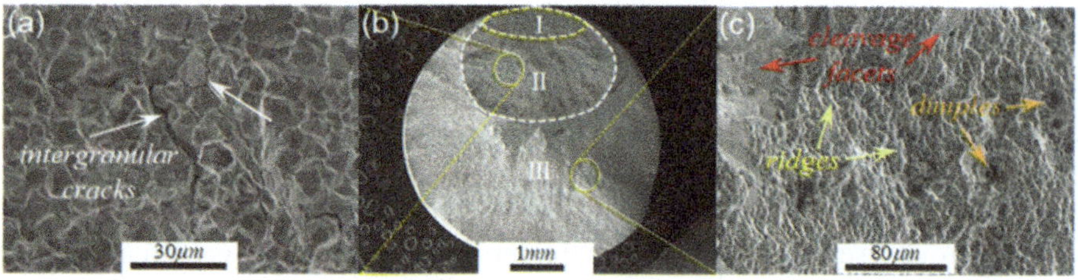

Fig. 7.3 (**a**) Intergranular cracks in zone II, (**b**) overview of the fracture surface, and (**c**) transient fracture zone showing signs of both ductile and brittle fracture [40]

7.4 Enhancing Creep Resistance

As the performance of aeroengine components is often limited by their creep properties, numerous studies have focused on enhancing creep performance. There are various ways to do this: through material composition, manufacturing methods, and post-processing. Modeling and predicting creep life is an important topic, which is also the subject of numerous studies and has been reviewed extensively [41].

Various microstructural features are beneficial for improving creep life, such as reducing the quantity of grain boundaries. Based on the well-established Hall-Petch relationship, grain boundaries can resist the movement of dislocation networks. However, in these conditions, creep resistance decreases in the diffusional flow-controlled creep regime, as observed in Ni-based superalloys [42]. Therefore, conventional manufacturing techniques aim to increase grain size, and single-crystal Ni-based superalloys have been widely applied to turbine blades and are accepted in aeroengines.

The addition of solute elements to strengthen alloys via the formation or dispersion of strengthening phases or through solid solution has been proposed and widely attempted. The interactions of solute atoms with edge dislocations lead to strong drag forces exerted during the climb processes [43]. Reis et al. [44] studied the creep behavior of a laser nitrided Ti-6Al-4 V alloy at 500 °C, at stress levels of 319 MPa to 520 MPa, and found that the laser nitrided sample had lesser oxidation effects, resulting in better creep life, due to the hard surface and interstitial solid solution. For Ni-based superalloys, a large number of alloying elements, including Co, Cr, Mo, W, Al, Ti, Ta, Re, and Ru, have been used to improve oxidation resistance and stabilize microstructures during long-term service [45]. For example, the addition of 3 wt.% Re to Ni-based single-crystal superalloys resulted in higher strength at elevated temperatures but also introduced some intermetallics, which act as void nucleation sites and decrease creep resistance [46]. Therefore, the addition of elements needs to be done with caution so that the strengthening elements outweigh potential negative counterparts.

Nonconventional manufacturing techniques, such as additive manufacturing (AM), can enable high manufacturing efficiency, complex designs, and reduce material waste. The use of AM, and more particularly of powder bed fusion, of Ni-based superalloys has been extensively studied [47]. Furthermore, numerous studies have observed comparable and even improved creep properties of Ni-based superalloys, as compared to conventional manufacturing methods [48]. Im et al. [49] and Pröbstle et al. [50] reported that AM-fabricated Inconel 718 (non-heat-treated) showed better creep resistance than that of conventional forged Inconel 718 with heat treatment, as can be seen from the example in Fig. 7.4a. However, AM of Ni-based superalloys results in a very different microstructure than conventionally manufactured parts (e.g., elongated grains and high texture in the <100> direction), as shown in Fig. 7.4b, c. The various AM process parameters, such as build orientation, scan strategies, and the use of multiple lasers, result in an anisotropic microstructure [51, 52] with porosity [53] and affect the creep performance [54]. The

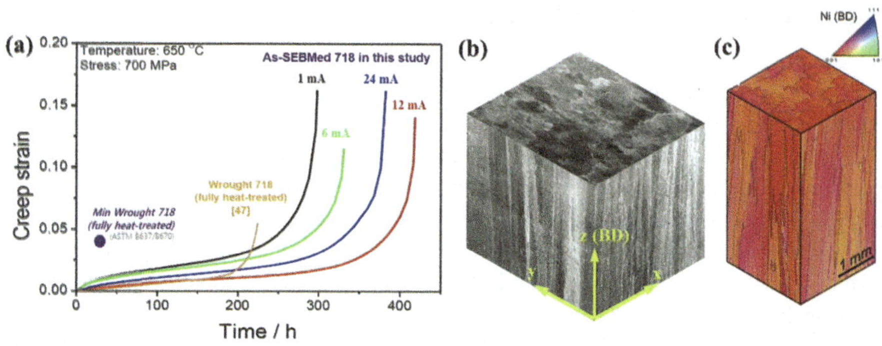

Fig. 7.4 (**a**) Comparison of creep curves of non-heat-treated Electron Beam Melted Inconel 718 samples, (**b**) microstructures of non-heat-treated Electron Beam Melted Inconel 718 samples (fabricated with the focus offset of 12 mA), and (**c**) Inverse Pole Figure (IPF) maps showing the building direction (BD) parallel with <001> crystallographic orientation from bottom to top [49]

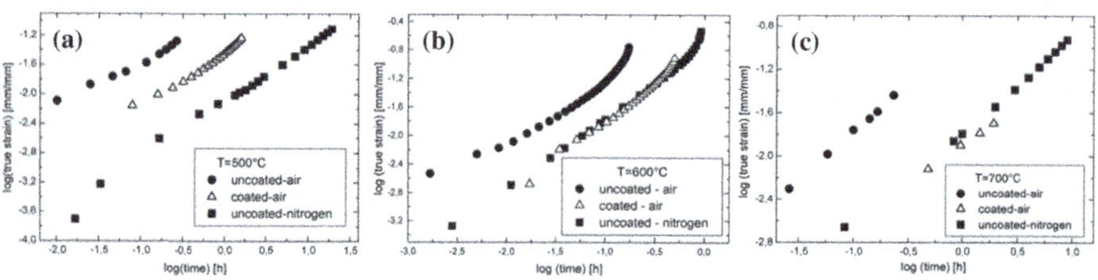

Fig. 7.5 The double logarithm plot of true strain against time for Ti-6Al-4 V at (**a**) 500 °C, (**b**) 600 °C, and (**c**) 700 °C [57]

higher porosity in AM materials is particularly detrimental to creep properties [55] as they act similarly to voids that can propagate.

Post-processing methods can also be used to enhance creep performance. Thermal barrier coatings (TBCs) have been considered by engine designers in recent years for thermal isolation of hot-section components through the reduction of the thermal expansion gradient between the substrate and the top coat. Kucuk et al. [56] have found that plasma-sprayed yttria (8 wt.%) stabilized zirconia (YSZ) can offer good thermal insulation from the metal material because of its porous microstructure and ceramic characteristics. Together with the NiCrAlY bond coat, which can provide better adhesion strength between the metal component and the YSZ coating, the application of the TBCs enables the enhancement of thermal resistance and oxidation resistance under high temperatures. This was confirmed by the research of Reis et al. [57], who found that the coating enhanced the creep resistance more than uncoated samples tested in air and that the efficiency of the coating increased with the applied measurement temperatures (Fig. 7.5). A two-layer coating system consisting of a 0.25–0.51 mm thermal-sprayed stabilized zirconia outer layer and a 0.10–0.20 mm MCrAlY inner layer (where M = Ni, Co, Fe, or combinations of these elements) has generally been the preferred configuration. It should be mentioned that the thickness of the coating may introduce residual stress [56], and thus, preheating is required before the coating process [58]. Another critical concern is the control of the porosity of the coating layer, which allows the formation of a thermally grown oxide layer. The overgrowth of this oxide layer can result in the failure of the TBC layer. The management and manipulation of the microstructure have been explored and achieved through alternative deposition processes and post-coating processes [59].

Heat treatment is another effective and widespread method to improve mechanical performance by eliminating harmful intermetallics, releasing residual stresses, homogenizing the microstructure, forming strengthening phases, and dissolving detrimental phases. HT can be classified into solution annealing, stress relieving, homogenization annealing, and/or aging for particular purposes. Du et al. [60] explored a low-density Ni-based single-crystal superalloy with a solution treatment at 1320 °C for different time periods, followed by primary aging at 1080 °C and secondary aging at 900 °C, which homogenized the microstructure and contributed enhanced creep properties. After the aging process, Kim et al. [61] found that the eutectic γ/γ' in cast Ni-based alloy, as shown in Fig. 7.6, acted as a barrier to prevent dislocation movement. Inversely, Gallmeyer et al. [62] performed solution annealing at 1020 °C for 15 min, followed by aging at 720 °C for 24 h to dissolve undesired Laves and δ phases in AM-fabricated IN718. Depending on the manufacturing

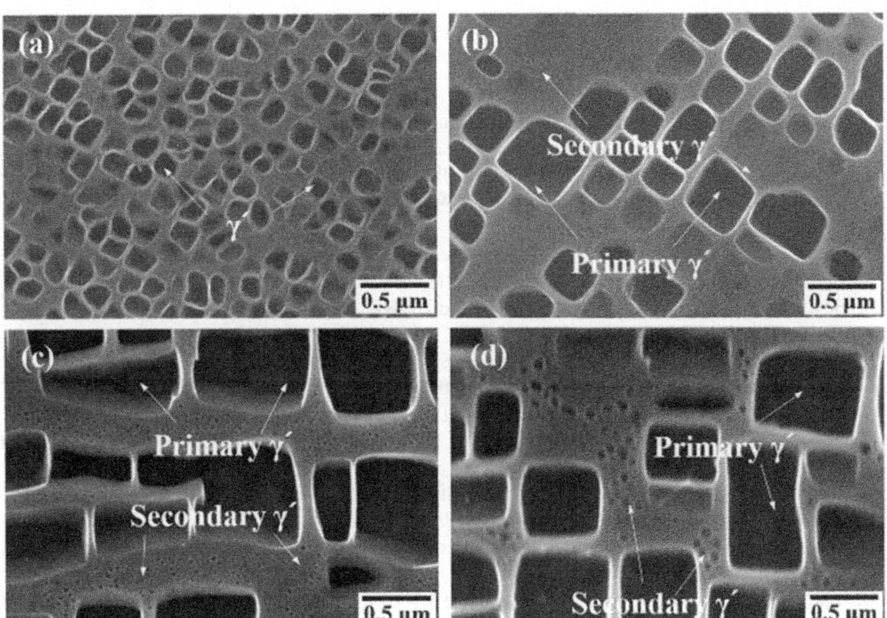

Fig. 7.6 Morphologies of γ' particles in heat-treated CM247LC: (**a**) 1232 °C/2 h/air cooling (AC), (**b**) 1232 °C/2 h/AC + 1080 °C/4 h/AC, (**c**) 1080 °C/4 h/AC, and (**d**) 1080 °C/4 h/AC + 871 °C/20 h/AC [61]

method employed, the heat treatment varies in terms of temperature and time. For example, wire and arc-based processes necessitate longer solution treatments at higher temperatures [63].

7.5 Conclusion

This chapter presents an overview of creep failure in aeroengine components and, more specifically, of Ni-based superalloys. The main creep failure mechanisms, the influence of temperature and load, as well as some highlights on enhancing creep performance, are discussed. It is noted that the creep failure mechanisms depend on the temperature and loading conditions and are controlled by diffusion and dislocation activity. At low temperatures, dislocation motion is restricted, and creep can only occur when high stresses are also present, whereas at high temperatures, creep can occur even at low stresses. Additionally, Ni-based superalloy phases have a significant effect on creep failure. Particularly, γ' is a key strengthening phase, which can coarsen during creep, contributing to failure. Finally, it is shown that various material formulations, manufacturing processes, and post-processing methods have the potential to enhance creep performance.

References

1. Rao Vakada N, Naga V, Rao B, Niranjan Kumar IN, Prasad KB, Madhulata N, et al. Failure mechanisms in turbine blades of a gas turbine Engine-an overview. Int J Eng Res. 2014;10:8–57.
2. Gasson PC. In: Reed RC, editor. The superalloys: fundamentals and applications. Cambridge: Cambridge University Press; 2006. 372 pp. Illustrated. £80. ISBN 0-521-85904-2. The Aeronautical Journal. 2008;112(1131).
3. Smith RJ, Lewis GJ, Yates DH. Development and application of nickel alloys in aerospace engineering. Aircr Eng Aerosp Technol. 2001;73(2):138.
4. Boyer RR, Cotton JD, Mohaghegh M, Schafrik RE. Materials considerations for aerospace applications. MRS Bull. 2015;40(12):1055.
5. Pollock TM, Tin S. Nickel-based superalloys for advanced turbine engines: chemistry, microstructure, and properties. J Propuls Power. 2006;22:361.
6. Miller S. Advanced materials mean advanced engines. Interdiscip Sci Rev. 1996;21(2):117.
7. Geddes B, Leon H, Huang X. Superalloys: alloying and performance. Materials Park: ASM International; 2010.
8. Choudhury IA, El-Baradie MA. Machinability of nickel-base super alloys: A general review. J Mater Process Technol. 1998;77(3–4):278–84.
9. Ezugwu EO, Wang ZM, Machado AR. The machinability of nickel-based alloys: A review. J Mater Process Technol. 1998;86(1–3):1–16.
10. Ezugwu EO, Bonney J, Yamane Y. An overview of the machinability of aeroengine alloys. J Mater Process Technol. 2003;134(2):233.
11. Zhang JX, Murakumo T, Koizumi Y, Kobayashi T, Harada H, Masaki S. Interfacial dislocation networks strengthening a fourth-generation single-crystal TMS-138 superalloy. Metall Mater Trans A Phys Metall Mater Sci. 2002;33(12):3741.
12. Blum W, Eisenlohr P. Dislocation mechanics of creep. Mater Sci Eng A. 2009;510-511:7–13.

13. Zhu H, Muránsky O, Wei T, Davis J, Budzakoska-Testone E, Huang H, et al. The effect of applied stress on the high-temperature creep behaviour and microstructure of NiMoCr Hastelloy-N® alloy. Materialia (Oxf). 2021;16:101069.
14. Ejaz N, Tauqir A. Failure due to structural degradation in turbine blades. Eng Fail Anal. 2006;13(3 Spec. Iss):452–63.
15. Ejaz N, Qureshi IN, Rizvi SA. Creep failure of low pressure turbine blade of an aircraft engine. Eng Fail Anal. 2011;18(6):1407.
16. Wang XM, Hui YZ, Hou YY, Yu ZY, Li L, Yue ZF, et al. Direct investigation on high temperature tensile and creep behavior at different regions of directional solidified cast turbine blades. Mech Mater. 2019;136:103068.
17. Sahoo B, Panigrahi SK, Satpathy RK. Creep life degradation and microstructure degeneration in a low-pressure turbine blade of a military aircraft engine. J Fail Anal Prev. 2017;17(3):529.
18. Karpinos BS, Samuleev VV, Lyashenko BA, Lais'ke EV, Rutkovs'kyi AV. Thermal cyclic creep and long-term strength of the material of aircraft gas turbine blades after operation. Strength Mater. 2013;45(5):542.
19. Xia W, Zhao X, Yue L, Zhang Z. Microstructural evolution and creep mechanisms in Ni-based single crystal superalloys: a review. J Alloys Compd. 2020;819:152954.
20. Unocic RR, Viswanathan GB, Sarosi PM, Karthikeyan S, Li J, Mills MJ. Mechanisms of creep deformation in polycrystalline Ni-base disk superalloys. Mater Sci Eng A. 2008;483–484:25–32.
21. Yao HT, Xuan FZ, Wang Z, Tu ST. A review of creep analysis and design under multi-axial stress states. Nucl Eng Des. 2007;237(18):1969.
22. Dyson BF, McLean M. Particle-coarsening, $\sigma 0$ and tertiary creep. Acta Metall. 1983;31(1):17.
23. Epishin A, Link T. Mechanisms of high-temperature creep of nickel-based superalloys under low applied stresses. Philos Mag. 2004;84(19):1979.
24. Murakumo T, Kobayashi T, Koizumi Y, Harada H. Creep behaviour of Ni-base single-crystal superalloys with various γ' volume fraction. Acta Mater. 2004;52(12):3737.
25. Ro Y, Koizumi Y, Harada H. High temperature tensile properties of a series of nickel-base superalloys on a γ/γ' tie line. Mater Sci Eng A. 1997;223(1–2):59–63.
26. Jahangiri MR, Abedini M. Effect of long time service exposure on microstructure and mechanical properties of gas turbine vanes made of IN939 alloy. Mater Des. 2014;64:588.
27. Sanchez S, Gaspard G, Hyde CJ, Ashcroft IA, Ravi GA, Clare AT. On the thermomechanical aging of LPBF alloy 718. Mater Sci Eng A. 2022;841:841.
28. Wollgramm P, Buck H, Neuking K, Parsa AB, Schuwalow S, Rogal J, et al. On the role of Re in the stress and temperature dependence of creep of Ni-base single crystal superalloys. Mater Sci Eng A. 2015;628:382–95.
29. Smith TM, Unocic RR, Deutchman H, Mills MJ. Creep deformation mechanism mapping in nickel base disk superalloys. Mater High Temp. 2016;33:372.
30. Wollgramm P, Bürger D, Parsa AB, Neuking K, Eggeler G. The effect of stress, temperature and loading direction on the creep behaviour of Ni-base single crystal superalloy miniature tensile specimens. Mater High Temp. 2016;33(4–5):346–60.
31. Wollgramm P, Wu X, Eggeler G. On the temperature dependence of creep behavior of Ni-base single crystal superalloys. In: Proceedings of the international symposium on superalloys. 2016.
32. Naeem M. Implications of day temperature variation for an aero-engine's HP turbine-blade's creep life-consumption. Aerosp Sci Technol. 2009;13(1):27.
33. Epishin AI, Fedelich B, Viguier B, Schriever S, Svetlov IL, Petrushin NV, et al. Creep of single-crystals of nickel-base γ-alloy at temperatures between 1150 °C and 1288 °C. Mater Sci Eng A. 2021;825:141880.
34. Nörtershäuser P, Frenzel J, Ludwig A, Neuking K, Eggeler G. The effect of cast microstructure and crystallography on rafting, dislocation plasticity and creep anisotropy of single crystal Ni-base superalloys. Mater Sci Eng A. 2015;626:305–12.

35. Heep L, Bürger D, Bonnekoh C, Wollgramm P, Dlouhy A, Eggeler G. The effect of deviations from precise [001] tensile direction on creep of Ni-base single crystal superalloys. Scr Mater. 2022;207:114274.
36. Agudo Jácome L, Nörtershäuser P, Heyer JK, Lahni A, Frenzel J, Dlouhy A, et al. High-temperature and low-stress creep anisotropy of single-crystal superalloys. Acta Mater. 2013;61(8):2926.
37. Agudo Jácome L, Nörtershäuser P, Somsen C, Dlouhý A, Eggeler G. On the nature of γ' phase cutting and its effect on high temperature and low stress creep anisotropy of Ni-base single crystal superalloys. Acta Mater. 2014;69:246.
38. Rae CMF, Reed RC. Primary creep in single crystal superalloys: origins, mechanisms and effects. Acta Mater. 2007;55(3):1067.
39. He DG, Lin YC, Yin LX. Creep characteristics and fracture mechanisms of a Ni-based Superalloy with δ phases at intermediate temperatures. Adv Eng Mater. 2020;22(8):2000144.
40. Zhao X, Zhang L, Song Y, Niu X, Sun Z, Zhao H, et al. Creep deformation and damage characteristics of nickel-based superalloy GH4169. Mater Sci Eng A. 2023;885:145589.
41. Yuan L, Chenglong W, Yongchao S, Mingbo S, Yu Y, Zhan G, et al. Review on creep phenomenon and its model in aircraft engines. Int J Aerosp Eng. 2023;2023:1.
42. Santella ML, Tortorelli PF, Render M, Pint B, Wang H, Cedro V, et al. Effects of applied stress and grain size on creep-rupture lifetime prediction for Haynes 282 alloy. Mater Sci Eng A. 2022;838:142785.
43. Farkoosh AR, Dunand DC, Seidman DN. Solute-induced strengthening during creep of an aged-hardened Al-Mn-Zr alloy. Acta Mater. 2021;219:117268.
44. dos Reis AG, Reis DAP, De Moura NC, Barboza MJR, Neto FP, Onõro J. Creep behavior study at 500°C of laser nitrided Ti-6Al-4 v alloy. J Mater Res Technol. 2013;2(1):48.
45. Long H, Mao S, Liu Y, Zhang Z, Han X. Microstructural and compositional design of Ni-based single crystalline superalloys — A review. J Alloys Compd. 2018;743:203.
46. Yu J, Sun X, Jin T, Zhao N, Guan H, Hu Z. Effect of Re on deformation and slip systems of a Ni base single-crystal superalloy. Mater Sci Eng A. 2007;458(1–2):39–43.
47. Sanchez S, Smith P, Xu Z, Gaspard G, Hyde CJ, Wits WW, et al. Powder bed fusion of nickel-based superalloys: A review. Int J Mach Tools Manuf. 2021;165:103729.
48. Sanchez S, Gaspard G, Hyde CJ, Ashcroft IA, Ravi GA, Clare AT. The creep behaviour of nickel alloy 718 manufactured by laser powder bed fusion. Mater Des. 2021;204:109647.
49. Im SY, Jun SY, Lee JW, Lee JH, Lee BS, Lee HJ, et al. Unidirectional columnar microstructure and its effect on the enhanced creep resistance of selective electron beam melted Inconel 718. J Alloys Compd. 2020;817:153320.
50. Pröbstle M, Neumeier S, Hopfenmüller J, Freund LP, Niendorf T, Schwarze D, et al. Superior creep strength of a nickel-based superalloy produced by selective laser melting. Mater Sci Eng A. 2016;674:299–307.
51. Qiu Z, Wu B, Zhu H, Wang Z, Hellier A, Ma Y, et al. Microstructure and mechanical properties of wire arc additively manufactured Hastelloy C276 alloy. Mater Des. 2020;195:109007.
52. Qiu Z, Wang Z, van Duin S, Wu B, Zhu H, Wexler D, et al. Chapter 9 – A review of challenges and optimization processing during additive manufacturing of trademarked Ni-Cr-based alloys. In: Gürgen S, Pruncu CI, editors. Modern manufacturing processes for aircraft materials [Internet]. Elsevier; 2024. p. 263–309. Available from: https://www.sciencedirect.com/science/article/pii/B9780323953184000094.
53. Liu S, Shin YC. Additive manufacturing of Ti6Al4V alloy: A review. Mater Des. 2019;164:107552.
54. Sanchez S, Hyde CJ, Ashcroft IA, Ravi GA, Clare AT. Multi-laser scan strategies for enhancing creep performance in LPBF. Addit Manuf. 2021;41:101948.
55. Goel S, Sittiho A, Charit I, Klement U, Joshi S. Effect of post-treatments under hot isostatic pressure on microstructural characteristics of EBM-built Alloy 718. Addit Manuf. 2019;28:727–37.

56. Kucuk A, Berndt CC, Senturk U, Lima RS, Lima CRC. Influence of plasma spray parameters on mechanical properties of yttria stabilized zirconia coatings. I: four point bend test. Mater Sci Eng A. 2000;284(1–2):29–40.
57. Reis DAP, Moura Neto C, Silva CRM, Barboza MJR, Piorino Neto F. Effect of coating on the creep behavior of the Ti-6Al-4V alloy. Mater Sci Eng A. 2008;486(1–2):421–6.
58. Kotzem D, Arold T, Niendorf T, Walther F. Influence of specimen position on the build platform on the mechanical properties of as-built direct aged electron beam melted Inconel 718 alloy. Mater Sci Eng A. 2020;772:138785.
59. Thakare JG, Pandey C, Mahapatra MM, Mulik RS. Thermal barrier coatings—a state of the art review. Met Mater Int. 2021;27:1947.
60. Du Y, Yang Y, Diao A, Li Y, Wang X, Zhou Y, et al. Effect of solution heat treatment on creep properties of a nickel-based single crystal superalloy. J Mater Res Technol. 2021;15:4702–13.
61. Kim IS, Choi BG, Jung JE, Do J, Seok WY, Lee YH, et al. Effect of heat treatment on microstructural evolution and creep behaviors of a conventionally cast nickel-based superalloy. Mater Charact. 2020;165:110378.
62. Gallmeyer TG, Moorthy S, Kappes BB, Mills MJ, Amin-Ahmadi B, Stebner AP. Knowledge of process-structure-property relationships to engineer better heat treatments for laser powder bed fusion additive manufactured Inconel 718. Addit Manuf. 2020;31:100977.
63. Seow CE, Coules HE, Wu G, Khan RHU, Xu X, Williams S. Wire + arc additively manufactured Inconel 718: effect of post-deposition heat treatments on microstructure and tensile properties. Mater Des. 2019;183:108157.

Index

S. Gürgen (ed.), *Failure in Aircraft Materials*, Sustainable Aviation,
https://doi.org/10.1007/978-3-031-65850-1

SPRINGER NATURE

GPSR Compliance

The European Union's (EU) General Product Safety Regulation (GPSR) is a set of rules that requires consumer products to be safe and our obligations to ensure this.

If you have any concerns about our products, you can contact us on ProductSafety@springernature.com

In case Publisher is established outside the EU, the EU authorized representative is:

Springer Nature Customer Service Center GmbH
Europaplatz 3
69115 Heidelberg, Germany

The manufacturer's authorised representative in the EU is Springer Nature Customer Service Centre GmbH, Europaplatz 3, 69115 Heidelberg, Germany. If you have any concerns regarding our products, please contact ProductSafety@springernature.com

Printed and bound by CPI Group (UK) Ltd, Croydon, CR0 4YY
01/10/2026
02225875-0006